MARITIME PREPOSITIONING FORCE (FUTURE) CAPABILITY ASSESSMENT

Planned and Alternative Structures

ROBERT W. BUTTON, JOHN GORDON IV
DICK HOFFMANN, JESSIE RIPOSO, PETER A. WILSON

Prepared for the Office of the Secretary of Defense
Approved for public release; distribution unlimited

NATIONAL DEFENSE RESEARCH INSTITUTE

The research described in this report was prepared for the Office of the Secretary of Defense (OSD). The research was conducted in the RAND National Defense Research Institute, a federally funded research and development center sponsored by the OSD, the Joint Staff, the Unified Combatant Commands, the Department of the Navy, the Marine Corps, the defense agencies, and the defense Intelligence Community under Contract W74V8H-06-C-0002.

Library of Congress Cataloging-in-Publication Data

Maritime prepositioning force (future) capability assessment : planned and alternative structures / Robert W. Button ... [et al.].

p. cm.

Includes bibliographical references.

ISBN 978-0-8330-4950-6 (pbk. : alk. paper)

1. Navy-yards and naval stations--United States. 2. Sea-power--United States. 3. United States. Navy--Operational readiness. 4. United States. Marine Corps--Operational readiness. 5. United States. Navy--Appropriations and expenditures. 6. United States. Marine Corps--Appropriations and expenditures. 7. United States. Navy--Planning. 8. United States. Marine Corps--Planning. I. Button, Robert.

VA69.M18 2010
359.4'11--dc22

2010007713

Published 2010 by the RAND Corporation
1776 Main Street, P.O. Box 2138, Santa Monica, CA 90407-2138
1200 South Hayes Street, Arlington, VA 22202-5050
4570 Fifth Avenue, Suite 600, Pittsburgh, PA 15213-2665
RAND URL: http://www.rand.org/
To order RAND documents or to obtain additional information, contact
Distribution Services: Telephone: (310) 451-7002;
Fax: (310) 451-6915; Email: order@rand.org

Preface

Navy–Marine Corps concepts for sea basing would accelerate deployment and employment of naval power-projection capabilities by using the flexibility and protection provided by the Sea Base while minimizing the presence of support forces ashore. The Maritime Prepositioning Force (Future) as planned will provide needed sea basing capabilities, but its cost has been a source of concern. Capabilities provided by reduced (and more-affordable) sea basing forces are explored here. Opportunities to preserve planned capabilities with significant cost savings or to retain adequate capabilities with greater cost savings are demonstrated across a variety of scenarios. A related 2007 study by the RAND Corporation examined Navy–Marine Corps Sea Basing concepts that might be useful to the Army.[1]

This study should be of interest to the Department of the Navy, the Department of the Army, the Office of the Secretary of Defense, and Congress.

This research was sponsored by the office of the Director, Cost Assessment and Program Evaluation (CAPE), Office of the Secretary of Defense (OSD) and conducted within the Acquisition and Technology Policy Center of the RAND National Defense Research Institute, a federally funded research and development center sponsored by the Office of the Secretary of Defense, the Joint Staff, the Unified Com-

[1] Robert W. Button, John Gordon IV, Jessie Riposo, Irv Blickstein, and Peter A. Wilson, *Warfighting and Logistics Support of Joint Forces from the Joint Sea Base*, Santa Monica, Calif.: RAND Corporation, MG-649-NAVY, 2007.

batant Commands, the Navy, the Marine Corps, the defense agencies, and the defense Intelligence Community.

For more information on RAND's Acquisition and Technology Policy Center, contact the Director, Philip Antón. He can be reached by email at atpc-director@rand.org; by phone at 310-393-0411, extension 7798; or by mail at the RAND Corporation, 1776 Main Street, Santa Monica, California 90407-2138. More information about RAND is available at www.rand.org.

Contents

Preface iii
Figures vii
Tables ix
Summary xi
Acknowledgments xvii
Abbreviations xix

CHAPTER ONE
Introduction and Objectives 1
Introduction 1
Study Objectives 2
Research Approach 2
Organization of This Report 3

CHAPTER TWO
Operational Concepts and Scenarios 5
Sea Basing Operational Concepts 5
Marine Corps Concepts 5
Operational Scenarios 8
Single-MEB Sustainment Scenario 9
Two-MEB Sustainment Scenario 10

CHAPTER THREE
Major Combat Operations 15
Sustaining a Single MEB in an MCO 15
Sustaining Two MEBs in an MCO 26

CHAPTER FOUR
Counterinsurgency Operations 33
The Nature of COIN Operations 33
Possible Roles of the Sea Base in COIN 34
Logistics Requirements in COIN Operations 37

CHAPTER FIVE
The MPF(F) as a Joint Special Operations Task Force Afloat Forward Staging Base 45
Introduction 45
The Nature of Joint Special Operations 46
Role of Sea Bases in Special Operations 47
Capacity 48
Security 49
Mobility 49
Limitations 50
USSOCOM's Strategic Objectives 51
Conclusions on the Viability of the Sea Base for SOF 52

CHAPTER SIX
Conclusions 53

APPENDIXES
A. Additional Cases 57
B. Maritime Prepositioning Force (Future) Description 63
C. MPF(F) MEB Sustainment Requirements 71
D. Model Description 73

Bibliography 77

Figures

2.1. Operational Scenario for Single-MEB Sustainment 9
2.2. Operational Scenario for Two-MEB Sustainment 10
2.3. Operational Scenario for Two-MEB Sustainment 11
3.1. Tons per Day Requirement and Lift Capacities in MEB Sustainment 17
3.2. Single-MEB Sustainment Using All LHA(R)/LHD Assets ... 20
3.3. Single-MEB Sustainment Using All MPF(F) Connectors 21
3.4. Single-MEB Sustainment With CH-53K Helicopters Replacing MV-22 Aircraft 22
3.5. Single-MEB Sustainment Without One LHA(R) 23
3.6. Single-MEB Sustainment Without One LHA(R) and CH-53K Helicopters Replacing MV-22 Aircraft 24
3.7. Single-MEB Sustainment Without Both LHA(R)s 25
3.8. Single-MEB Sustainment Using LCACs from Three or Four MLPs 26
3.9. Air-Only Sustainment of Further MEB Not Feasible 28
3.10. Two-MEB Sustainment by Surface and Air Possible 28
3.11. Two-MEB Sustainment Without One MPF(F) LHA(R) Possible 29
3.12. Two-MEB Sustainment Without One LHA(R) and CH-53K Helicopters Replacing MV-22 Aircraft 30
3.13. Two-MEB Sustainment Without Both MPF(F) LHA(R)s 31
3.14. Two-MEB Sustainment Without Both MPF(F) LHA(R)s and LHD 31
4.1. Notional MEB Supply Requirements in COIN 38
4.2. Air-Only Resupply of a MEB in COIN 39
4.3. LHD LCACs and Aircraft Used to Support MEB in COIN 39

4.4. All MPF(F) Aircraft and LCACs Resupply a MEB in COIN 40
4.5. MEB in COIN Supported by MPF(F) Minus One LHA(R) 41
4.6. MEB Supported by MPF(F)—Minus One LHA(R)—With CH-53Ks Substituted for MV-22s 42
4.7. MEB in COIN Supported by MLP LCACs Only 43
A.1. Air-Only Sustainment of Further MEB Nearly Feasible With CH-53K Helicopters Replacing MV-22 Aircraft 58
A.2. Single MEB Supported by All or 70 Percent of MPF(F) Aerial Connectors Plus LHD LCACs 59
A.3. Single MEB Supported by All or 70 Percent of MPF(F) Aerial Connectors 60
A.4. Single MEB With Ammunition Movement by Air Connectors Only 61
B.1. LHD-5, USS *Bataan* 65
B.2. T-AKE-1, USNS *Lewis and Clark* 66
B.3. T-AKR-317, USNS *Soderman* 67
B.4. Vehicle Transfer Between MPF(F) LMSR and Conceptual Mobile Landing Platform 68
B.5. Conceptual MLP 69
D.1. Lift Capacity Using LHA(R)/LHD Assets in MEB Sustainment 76

Tables

2.1. Percentage of Sea State 3 or Less Conditions for Various Littoral Regions 8
C.1. SBME Daily Sustainment Requirements 71

Summary

This study examines various possible changes to the planned composition of the Maritime Prepositioning Force (Future) (MPF[F]). At the time of this analysis, the planned 14-ship MPF(F) squadron will consist of

- two modified LHA (replacement) (LHA[R]) large-deck amphibious assault ships equipped with Marine Expeditionary Brigade command and control facilities
- one modified LHD large-deck amphibious ship
- three *Lewis and Clark* dry cargo/ammunition ships (T-AKEs)
- three modified large, medium-speed, roll-on/roll-off (LMSR) sealift ships
- three mobile landing platform (MLP) ships each capable of operating six landing craft, air cushioned (LCAC) surface connectors
- two legacy "dense-pack" MPF ships taken from existing squadrons.[2]

LHA(R)s and LHDs have large flight decks and hangar decks for embarking and operating helicopters and tilt-rotor aircraft. LHA(R)s and LHDs also provide medical capabilities: With six operating rooms, 17 intensive care unit beds and 60 overflow beds, LHDs have the greatest medical capability of any amphibious platform in operation.

---

[2] A more detailed description of the program of record MPF(F) squadron is provided in Appendix B.

Most of the possible variations from the program of record MPF(F) entail removing large-deck ships: the LHA(R)s and the LHD. Additionally, we examined cases where a fourth MLP is added (with six additional LCACs) in order to assess a situation where only surface connectors can be used. Most of our analysis focuses on major combat operations (MCOs), but we did consider the possibility of the MPF(F) supporting counterinsurgency (COIN) and special operations. This analysis does not examine the ability of the MPF(F) to support joint operations; rather, it concentrates on the ability of a modified MPF(F) to sustain United States Marine Corps (USMC) Marine Expeditionary Brigades (MEBs).

Analysis and Scenarios

Our analysis assessed potential sea base logistic support performance in MCOs and in COIN operations. It also treated potential roles and capabilities for MPF(F) as an afloat forward staging base for joint special operations. Our MCO scenario cases included support to a single MEB and simultaneous support to two MEBs. Our MCO scenario cases also included scenarios in which constraints (such as a requirement for air-only sustainment) are placed on the use of MPF(F) connectors. We varied MPF(F) assets within the context of the MCO and COIN operation cases. Sustainment distances were also varied for each case. The result is a wide-ranging logistic support analysis.

Our analysis was not limited to the consideration of logistic support. We also considered implications for casualty evacuation (CASEVAC) and care and for the movement of supplies and equipment ashore. In that regard, note that the planned composition of the MPF(F) is based in large part on the USMC requirement that the ground-maneuver elements of the MEB that is carried aboard the MPF(F) be capable of moving ashore in one period of darkness. That requirement to a large extent drives the need for the 18 LCAC hovercraft aboard the new MLP ships (or 21 LCACs including those aboard the single LHD in the squadron). The relatively large tonnage capacity of the LCACs is needed to deploy several thousand Marines and their equipment ashore in one cycle of darkness. Once the maneuver elements are ashore, however, the daily tonnage requirement of the MEB is far less than the theoretical throughput from ship-to-shore that the LCACs are capable of —not even including the aircraft aboard the squadron. Readers should keep that reality in mind as they encounter study results that show, in many cases, a large theoretical sustainment "excess capacity."

Key Findings

- *Eliminating one LHA(R).* The degradation to logistics throughput resulting from the elimination of one LHA(R) could be offset in all cases by substituting CH-53K helicopters for MV-22 tilt-rotor aircraft; CH-53K helicopters have three times the payload of the MV-22 and, in our scenarios, are just as fast on ingress.[3] The MV-22's higher speed is advantageous in CASEVAC operations, where time is critical and external loads do not limit its speed.
- *Eliminating both LHA(R)s.* The degradation to logistics throughput resulting from eliminating both LHA(R)s cannot be offset by substituting CH-53K helicopters for MV-22 tilt-rotor aircraft; too few aircraft then remain in the MPF(F) squadron. However, a robust throughput capacity remains for all cases considered using air connectors from the remaining LHD and LCACs from the LHD and the MLPs.
- *Eliminating all large decks.* The elimination of all three large decks (both LHA[R]s and the LHD) in the MPF(F), with sustainment conducted entirely using LCACs from MLPs, leaves a marginal capacity to sustain a single MEB (either in MCO or in COIN operations) with three or four MLPs. However, this option also strips out the MPF(F) squadron's major medical capabilities and

[3] Both helicopters and tilt-rotor aircraft are expected to carry external loads in sustainment operations. The advantage of greater payload weight for internal loads is more than offset by the additional loading time required for external loads. On ingress, aerodynamic constraints imposed by external loads make the MV-22 no faster than the CH-53K. The MV-22 can employ its high speed only on returning from the shore to the sea base.

forces a reliance on slower aircraft for CASEVAC. Aviation command and control capabilities provided by the LHD would also be lost. Further, the ability of MLPs to work with T-AKEs could be constrained by the relatively small number of helicopters carried for vertical replenishments (VERTREPs) by T-AKEs. Finally, removing the aircraft associated with the large flight decks could impose tactical constraints on the MEB commander. While our analysis showed that, in terms of raw throughput, the LCACs of three or four MLPs could meet the MEB's daily tonnage requirements, we found that significant operational issues could arise without air-capable ships in the force. When sustaining two MEBs in conjunction with an amphibious task force (ATF), throughput capacity is marginal without the addition of a fourth MLP. However, the loss of medical and CASEVAC capabilities is less of an issue. The issues of ability of the MLPs to work with T-AKEs and constraints on the MEB commander are also mitigated by the presence of an ATF.

- *LCAC capacity.* The bulk of ship-to-shore throughput capacity for MPF(F) connectors resides with LCACs. The combined total throughput capacity of the 21 LCACs carried by the MPF(F) alone significantly exceeds daily tonnage requirements for the 2015 MEB. Moreover, the Marine Corps Combat Devleopment Command's (MCCDC's) sustainment plans use the three LCACs carried by the LHD; the use of LCACs cannot be discounted completely. However, as noted below, we uncovered important issues associated with a heavy reliance upon LCACs for sustainment.
- *Ashore connectors.* Supplies delivered to the shore by LCACs must be moved forward from the beach (or small port) to the USMC or other forces that will consume the supplies. Such movement requires a quantity of trucks and/or aircraft and a reasonably secure area through which they can move. These conditions will, of course, be situationally dependent.
- *T-AKE/MLP interface.* The USMC does not currently envision a direct T-AKE/MLP interface; the offloading of supplies from the T-AKEs is presently limited to aircraft-only sustainment. This concept would shift the burden of lift from LCACs to vertical-lift

connectors and so reduce the number of CH-53K and MV-22 sorties available to joint force commanders for purposes other than sustainment.[4] In order to realize the full potential of the MLPs' LCACs, we recommend the Navy and USMC investigate ways that the T-AKEs could interface more closely with the LCACs, either by directly offloading onto the hovercraft themselves or by transloading supplies from T-AKE to MLP, and then into the LCACs.

- *Other LCAC missions.* If the Marine Corps cannot use the full potential of the LCACs, the joint force commander should consider ways to use LCACs for movement ashore and sustainment of other forces. For example, if the MEB does not need, or cannot make use of, the LCACs' throughput potential, the Army could offload personnel, supplies, and equipment onto the MLPs from Army LMSRs for movement ashore via LCAC.
- *Support for COIN.* The MPF(F) sea base, or portions of it, could provide important capabilities to support COIN operations. Although the daily tonnage requirements of a MEB engaged in COIN operations are situationally dependent, they would be lower than the consumption rates envisioned for MCO, especially in terms of ammunition. Therefore, the overall logistics throughput potential of the MPF(F) could easily support a MEB engaged in COIN operations, as well as additional USMC units or elements from the other services. Given the general desire that local forces have a leading role in COIN, the MPF(F) might also be used to support foreign forces. Finally, COIN operations might not require the employment of all the ships of the MPF(F), depending on the size and duration of the mission.
- *Support for Special Operations Forces (SOF).* The MPF(F) could provide a useful base for SOF operations. Even more than in support of COIN operations, support to SOF might require only a portion of the MPF(F). For example, a single MLP or an MLP plus a large flight deck from the MPF(F) might be sufficient to meet the needs of a SOF element, possibly for a protracted period of time.

---

[4] This topic is explored in some detail in Appendix A.

Acknowledgments

The authors would like to thank the sponsor of the study, Charles Werchado of OSD CAPE and his staff, who were always available for consultation and showed a keen interest in the results. Col. Joseph G. Smith, James Strock, and William A. Sawyers of the Marine Corps Combat Development Command offered frank and useful comments on the emerging study results. CAPT George Sutton and Jitendra Tandon, of NAVSEA, provided valuable updates on planned systems. In OPNAV N81, LCDR Jeremy Hill provided important inputs.

We wish to thank our reviewers, John Friel and Robert Murphy, for their feedback. Finally, the project assistant, Jennifer Miller, was very helpful in producing the final report.

Abbreviations

AFSB	afloat forward staging base
AoA	analysis of alternatives
APOD	aerial port of debarkation
ATF	Amphibious Task Force
BCT	brigade combat team
C2	command and control
CAPE	Cost Assessment and Program Evaluation
CASEVAC	casualty evacuation
CDD	capability development document
CJSOTF	Combined Joint Special Operations Task Force
CNA	Center for Naval Analyses
COIN	counterinsurgency
CONOPs	concept of operations
CONREP	connected replenishment
DoD	Department of Defense
ESG	Expeditionary Strike Group
FSB	forward staging base

JESOF	joint expeditionary SOF
JHSV	Joint High-Speed Vessel
JSF	Joint Strike Fighter
JSLM	Joint Seabasing Logistics Model
JSOAC	joint special operations air component
JSOTF	joint special operations task force
LCAC	landing craft, air cushioned
LCU	landing craft utility
LHA	amphibious assault ship, general purpose
LHA(R)	LHA (replacement)
LHD	amphibious assault ship, multipurpose
LMSR	large, medium-speed, roll-on/roll-off ship
MAGTF	Marine Air-Ground Task Force
MCCDC	Marine Corps Combat Development Command
MCO	major combat operation
MEB	Marine Expeditionary Brigade
MEU	Marine Expeditionary Unit
MLP	mobile landing platform
MPF	Maritime Prepositioning Force
MPF(F)	Maritime Prepositioning Force (Future)
MPSRON	Maritime Prepositioning Ship Squadron
NEO	noncombatant evacuation
NM	nautical mile
NSWTG	Naval Special Warfare Task Group

OPNAV	Office of the Chief of Naval Operations
OSD	Office of the Secretary of Defense
PEO	program executive office
POL	petroleum, oil and lubricants
SBE	Sea Base Echelon
SBME	Sea Base Maneuver Element
SBSE	Sea Base Support Element
SOF	special operations forces
SPOD	seaport of debarkation
SSBN	fleet ballistic missile submarine
SSGN	guided missile submarine
T-AKE	auxiliary dry cargo/ammunition ship
UAV	unmanned aerial vehicle
UNREP	heavy underway replenishment system
USMC	United States Marine Corps
USSOCOM	United States Special Operations Command
VERTREP	vertical replenishment

CHAPTER ONE

Introduction and Objectives

Introduction

The Department of Defense (DoD) is investing significant resources in developing seabasing capabilities that will allow the rapid deployment, assembly, command, projection, reconstitution, and re-employment of expeditionary forces from the sea. Obtaining these capabilities is an important part of achieving Sea Power 21, the Navy's operational vision for the 21st century. The Navy has proposed a new maritime prepositioned squadron, the Maritime Prepositioning Force (Future) or MPF(F), along with associated concepts of operations (CONOPs), to provide this capability as a component of its Sea Basing concept. Given the multiple pressures on the Navy's ship building budget, however, the MPF(F) as currently proposed may be unaffordable.[1] The residual capabilities provided by reduced (more affordable) Sea Basing forces need to be examined. Specifically, how would changing the number and mix of ships in the squadron affect the capability of the MPF(F) to support U.S. Marine Corps (USMC) forces? Such analyses are needed to ascertain the capability of the MPF(F) both when it is operating in isolation and in situations when it is operating together with other expeditionary forces.

---

[1] Ronald O'Rourke, *Navy-Marine Corps Amphibious and Maritime Prepositioning Ship Programs: Background and Oversight Issues for Congress,* Congressional Research Service, RL32513, updated July 10, 2007.

Study Objectives

The office of the Director, Cost Assessment and Program Evaluation, Office of the Secretary of Defense—OSD CAPE—and the Office of the Undersecretary of Defense for Acquisitions, Technology and Logistics—OUSD (AT&L)—asked the RAND Corporation to perform an independent assessment of various aspects of the Department of the Navy's plans to develop future Sea Basing capabilities. The objective of this research is to provide analysis to help OSD and the Department of the Navy to better understand the Sea Basing options associated with the 30-year shipbuilding plan. Specifically, this research assesses the current program of record and alternatives to the program of record for the proposed sea-based expeditionary forces for major combat operations (MCO), counterinsurgency operations (COIN), and special operations support.

Research Approach

This study built upon tools and knowledge gained through previous RAND research performed under the auspices of the National Defense Research Institute for the Assessment Division of the Office of the Chief of Naval Operations (OPNAV N81).[2] The RAND study team began by collecting data from the Navy and Marine Corps on consumption rates of supplies needed to support a Marine Expeditionary Brigade (MEB), associated operational concepts, and expected future force structure. We then developed operational vignettes representing MCO and COIN operations to evaluate alternative MPF(F) squadron structures. Under the MCO and COIN vignettes, RAND analysts constructed cases varying forces ashore, the number of ships in the MPF(F), and connectors (the aircraft and vessels used to move supplies ashore from the MPF[F]). We employed a simulation model to evaluate the ability of the various constructs to sustain forces ashore in these

---

[2] See Robert W. Button, John Gordon IV, Jessie Riposo, Irv Blickstein, and Peter A. Wilson, *Warfighting and Logistic Support of Joint Forces from the Joint Sea Base*, Santa Monica, Calif.: RAND Corporation, MG-649-NAVY, 2007.

cases. The RAND team also conducted a conceptual evaluation of how the MPF(F) and amphibious ships could be used to support special operations forces (SOF) operations.

Organization of This Report

Chapter Two describes the operational concepts and scenarios used to evaluate variations on the MPF(F) under a wide range of conditions. Chapter Three provides results for MCOs in situations where one or two MEBs must be supported. Chapter Four provides insights on the MPF(F)'s ability to support counterinsurgency operations. Chapter Five describes the applicability of MPF(F) assets in support of SOF. Chapter Six provides conclusions drawn from the previous chapters. Appendix A describes results from additional cases omitted from the main body of the report for brevity. Appendix B provides detailed descriptions of the planned MPF(F). Appendix C describes the MPF(F) MEB, the force to be sustained in the analysis. Appendix D describes the simulation model used for this study.

CHAPTER TWO

Operational Concepts and Scenarios

Sea Basing Operational Concepts

This section examines conceptual issues identified as part of this study. After highlighting key elements of Marine Corps concepts regarding use of the sea base, specifically the MPF(F), we introduce the operational scenarios used in the analysis.

Marine Corps Concepts

The Marine Corps regards the MPF(F) as a major step forward in their ability to operate from the sea under the rubric of *Operational Maneuver from the Sea*. Today's Maritime Prepositioning Ship Squadrons (MPSRONs) require safe, usable ports in which to offload cargo. Additionally, today's MPSRON ships are loaded in a *dense pack* configuration; that is, they are loaded to the maximum extent possible. Maximizing the cargo-carrying capacity of a ship increases the difficulty of offloading cargo, which means that several days of work at or near the seaport of debarkation (SPOD) are required before the MEB equipment carried onboard the MPSRON is operational.

MEBs can currently be brought into action in two ways: (1) via amphibious assault shipping (so-called "gray hulled" Navy amphibious vessels built to warship standards) or (2) the MPSRON (ships built to commercial standards that require a secure port to offload). The MPF(F) will provide a new option for movement ashore and the subsequent logistical support of a MEB. The USMC has determined that the MPF(F) squadron is not capable of amphibious assault. Instead, the Marines envision the MPF(F) squadron as an immediate follow-on to the assault MEB that will initially seize a lodgment. Significantly, whereas the current MPSRON needs a port to offload, the Marines and their equipment aboard MPF(F) ships will be able to conduct operations from offshore.

While Marine Expeditionary Units (MEUs) can deploy and sustain from their three-ship Expeditionary Strike Groups (ESGs), the MEU is a battalion-sized task force. The planned 14-ship MPF(F) squadron will give the Marines the ability to deploy and sustain an entire brigade (less its fixed-wing fighters).

Discussions with Marine Corps Combat Development Command (MCCDC) personnel revealed that the Marine Corps prefers to provide logistics support to the MEB once it is ashore via cargo-carrying aircraft (MV-22 tilt-rotor aircraft and CH-53K heavy-lift helicopters). This allows the MEB to (1) avoid creating a traditional "iron mountain" of shipborne supplies and material on the shore and (2) facilitates the rapid maneuver of the MEB inland once it is ashore. Further, the USMC wants to retain several MV-22s on the sea base for casualty evacuation (we accordingly set aside these MV-22s and associated deck spots in our analysis). Also, the Marine Corps envisions that some of the available MV-22 sorties (and possibly some of the CH-53K sorties) would be used for tactical mobility missions for the forces ashore.[1] For example, the MEB commander might want to conduct air assaults by company- or battalion-sized forces based on the tactical situation using some of the aircraft aboard the MPF(F) ships. In our analysis, the identification of excess air sorties (either MV-22 and/or CH-53K) could be interpreted as the ability (or not) of the sea base to simultaneously provide logistics support to USMC and U.S. Army forces ashore while at the same time retaining for the MEB commander the capability to conduct other maneuver-related air missions. Finally, the Marine Corps indicated a preference for air-only sustainment of ammunition supplies.

Current Navy plans envision the replacement of one of the three existing MPSRONs by a single MPF(F) squadron, which would prob-

[1] These preferences are reflected in our analysis as rules and data. The use of aircraft sorties for tactical mobility is explored in Appendix A.

ably be located at U.S. Naval Base Guam. In a future crisis requiring the deployment of multiple brigades, it is likely that a combination of ESGs, several of which are always on station at various locations around the world, and the MPF(F) squadron would form the initial USMC force. If a third MEB-sized force is needed, additional Marines aboard amphibious ships and/or the traditional dense packed MPSRON would arrive later to bring the USMC force ashore to division, or larger, size, if needed.

The Marine Corps envisions operating many of the MEB's aircraft from the sea base. However, the three large flight decks of the planned MPF(F) squadron are not sufficient to allow the Joint Strike Fighters (JSFs) of the MEB's air element to conduct sustained operations from the sea base (small numbers of JSFs could still use the MPF[F] as a base for refueling and rearming or emergency landings). It should also be noted that there would be little, if any, space available aboard the three flight decks of the MPF(F) for Army aircraft until and unless some portion of the MEB's MV-22s and helicopters is moved ashore or to another Navy vessel.

Several technical challenges are inherent in the MPF(F) concept. Perhaps the most critical challenge is the difficulty of ship-to-ship transfer in high sea states. Current planning would restrict ship-to-ship transfers in conditions greater than sea state 3 on the Pierson-Moskowitz sea state scale. The implications of this limitation vary globally. Table 2.1 shows the frequency of occurrence for conditions of sea state 3 or less over various regions on an annual basis.[2]

This table indicates that MPF(F) operations will not be degraded by high sea states at least 70 percent of the time in the high-profile regions of the Persian Gulf and the North Arabian Sea, the Mediterranean Sea, the Gulf of Guinea, and the Korean Coast. However, stated less positively, MPF(F) operations in these regions *will* be degraded up to 30 percent of the time. It would therefore be prudent to provide a significant reserve throughput capacity to compensate for heightened sea states and other real-world considerations. In this regard, a capacity roughly 50 percent greater than required throughput requirements

---

[2] Defense Science Board, *Enabling Sea Basing Capabilities*, August 2003, p. 37.

Table 2.1
Percentage of Sea State 3 or Less Conditions for Various Littoral Regions

Western Atlantic	60	Mediterranean Sea	75
Eastern Atlantic	40	Persian Gulf	89
North Sea (including English Channel)	52	North Arabian Sea	73
Eastern Pacific	45	West Indian Ocean	52
West and South Caribbean	53	Cape of Good Hope	21
Northeastern South America	54	Gulf of Guinea	71
Western South Atlantic	43	North West Africa	48
Eastern South Pacific	40	East Coast of Japan	48
Northwestern South America	55	East Coast of the Philippines	62
Western Central America	73	Korean Coast	71

appears necessary for these high-profile regions to enable reliable sustainment for protracted periods of time.[3]

Operational Scenarios

We developed two broad operational scenarios for this analysis: a scenario in which a single MEB is sustained from the sea base and a scenario in which two MEBs are sustained from the sea base. We associated multiple cases with each scenario. For example, there are cases in which a single MEB is engaged in an MCO and cases in which a single MEB is engaged in a COIN operation. Multiple cases arise with differing formulations of the MPF(F) squadron, differing mixes of connectors, and restrictions on connector usage (such as the constraint that ammunition can only be transported by air). Unlike our 2007 analysis of the MPF(F) that focused on how it might be able

[3] With a capacity 150 percent of the required throughput capacity, the ability to operate at full capacity 70 percent of the time results in a residual capacity closely matching the required throughput capacity ($1.5 \times 0.7 \sim 1.0$).

to simultaneously support both Army and USMC forces,[4] this study focuses on the ability of the MPF(F) to support one or two Marine Corps MEBs (as well as SOF forces where relevant in the analysis).

Single-MEB Sustainment Scenario

The single-MEB sustainment scenario (shown in Figure 2.1) entails sustaining a single MEB in MCO or COIN operations using all or some MPF(F) assets. MLPs are assumed to operate approximately 25 miles offshore, with the distance to the center of mass of the MEB located 25 to 110 NM from the MPF(F)'s large flight decks.

The SPOD in this scenario is envisaged to be a minor port or perhaps a beach. Landing craft, air cushioned (LCAC) connectors (Navy-operated hovercraft that can load personnel, supplies, and equipment at the sea base and deliver them to the beach) are launched at a distance of 25 NM from the SPOD from amphibious assault ships, general pur-

Figure 2.1
Operational Scenario for Single-MEB Sustainment

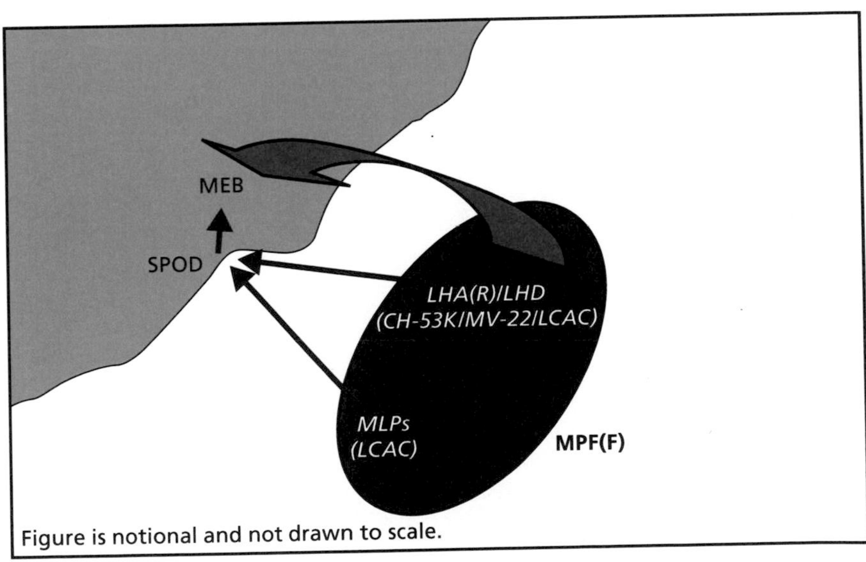

RAND *MG943-2.1*

[4] Button et al., 2007.

pose (replacement) (LHA[R]s); amphibious assault ships, multipurpose (LHDs); or the new mobile landing platforms (MLPs). LCACs can operate at speeds of over 30 knots and can carry up to 70 tons of payload in each load. In this scenario, aircraft are launched at distances of 25 to 110 NM from the MEB (depending upon the brigade's distance inland or along the coastline).

Two-MEB Sustainment Scenario

We looked at three variants of the scenario for sustaining two MEBs simultaneously:

- Both MEBs are sustained by air only (using various aircraft mixes) at distances of 25 to 110 NM.
- The nearer MEB is sustained by LCACs operating 25 NM from the SPOD with air sustainment at distances of 25 to 110 NM, while the further MEB is sustained by aircraft from distances of 75 to 160 NM (Figure 2.2).

Figure 2.2
Operational Scenario for Two-MEB Sustainment (further MEB sustained by air only)

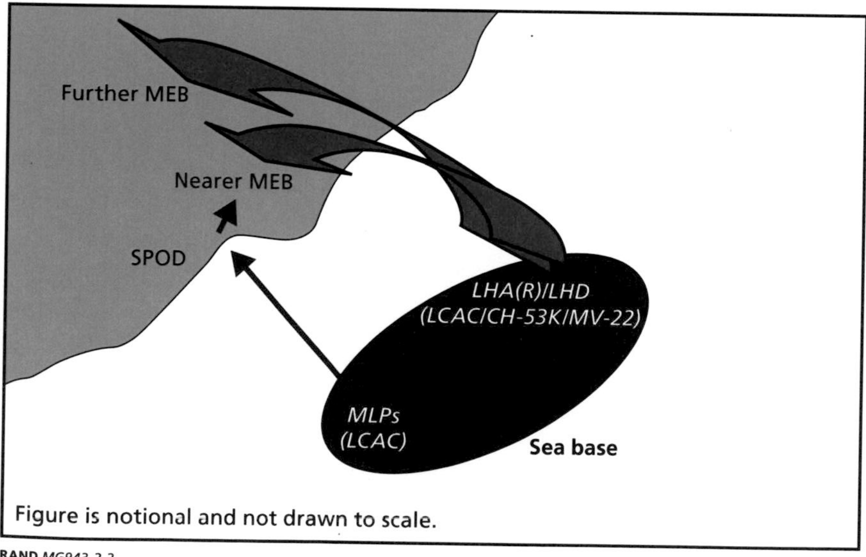

RAND *MG943-2.2*

- Both MEBs are sustained by LCACs (operating 25 NM from the SPOD) and by aircraft (operating at distances of 25 to 110 NM from the nearer MEB and 75 to 160 NM from the further MEB) (Figure 2.3).

The use of LCACs as part of resupply efforts is a key factor examined in these scenarios. As observed earlier, the Marine Corps prefers that, once ashore, the MEB is resupplied to the maximum extent possible by aircraft operating from the sea base. In cases where LCACs are used, we assumed that either (1) the MEBs are close enough to the coast so that it would be easy to retrieve supplies delivered to the beach by LCACs or (2) the units were fairly deep inland (25 miles or more) but had the ability to dispatch trucks to the beach to retrieve the supplies delivered by LCACs.[5]

Figure 2.3
Operational Scenario for Two-MEB Sustainment (both MEBs sustained by surface and air)

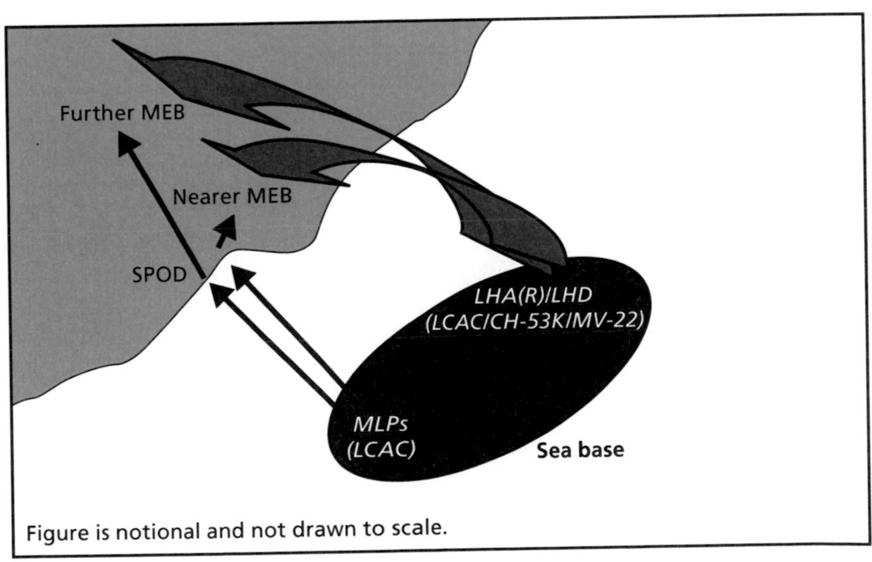

RAND *MG943-2.3*

5 See Appendix C for additional information.

It is important to note that the planned composition of the MPF(F) is based in large part on the USMC requirement that the ground maneuver elements of the MEB carried aboard the MPF(F) be capable of moving ashore in one period of darkness. To a large extent, that requirement drives the need for the 18 hovercraft (or 21 if the LCACs aboard the single LHD in the squadron are included) aboard the new MLP ships. The relatively large tonnage capacity of the LCACs is needed to deploy several thousand Marines and their equipment ashore in one 8- to 10-hour period of darkness. Once the maneuver elements are ashore, however, the daily tonnage requirement of the MEB is far less than the ship-to-shore throughput capacity provided by LCACs alone—not even including the aircraft aboard the squadron. Readers should keep this reality in mind as they review the results of our study, which in many cases show a large theoretical excess of sustainment capacity.

The assumption that the MEBs could retrieve LCAC-delivered supplies for movement inland by ground transport presumes that the routes from the beach (or small port) to the units operating inland are relatively safe. This may not always be the case, thus requiring the ground units to escort their supply vehicles and provide protection for the offload points at the beach or port. It should be noted that we did not envision a large amount of infrastructure being built to support operations at the beach, certainly nothing like the iron mountains associated with World War II–type amphibious operations. Sustainment operations would instead only maintain several days of supplies ashore. Nevertheless, the MEB commanders and any Joint Force commander would have to accept the implications of cross-beach supply. A downside to this situation is the possibility of needing to provide protection and escort for the supplies arriving at and moving forward from the beach. However, the advantage is that if LCACs are used to supplement the aerial delivery of supplies from the MPF(F), the amount of tonnage that could be moved is increased substantially. It should also be noted that even if aerial resupply alone is being used—and the area between the shoreline and the units operating inland is not completely secured—the resupply aircraft would also be vulnerable to enemy fire as they pass over the unsecured area en route to deposit their supplies at inland locations. Finally, it should be noted that we did not analyze the number of trucks that would be required for the forward movement of supplies delivered to the beach by LCACs. We assumed that sufficient numbers of supply trucks (including trailers) would be available to the MEBs.

Another important consideration is the issue of how the Marine Corps envisions placing supplies aboard the ships of the MPF(F). The USMC envisions initially using the 21 LCACs in the MPF(F) to move a portion of the MEB ashore. As observed above, once the MEB is ashore, current Marine Corps concepts envision most of the supplies being flown from the ships of the MPF(F) to drop-off points ashore, if possible. Many of the supplies in the MPF(F) ships (especially ammunition and dry stores) are currently planned to be stored on the three dry cargo/ammunition (T-AKE) ships. Other types of supplies are carried aboard other ships of the MPF(F), such as the LHD and the large, medium-speed, roll-on/roll-off (LMSR) ships that are configured for easy interface with MLPs. T-AKEs do not normally interface directly with other ships (or LCACs). Instead, T-AKEs have a small flight deck for use by embarked helicopters or by aircraft from other ships from the MPF(F).[6] In addition to their vertical replenishment (VERTREP) capability, T-AKEs have an extensive connected replenishment (CONREP) capability; the *Lewis and Clark* class developed for the MPF(F) can simultaneously operate five CONREP stations or three CONREP stations while conducting VERTREP operations.[7]

In order to maximize the throughput potential from the MPF(F) to units ashore, the ability of the T-AKE to directly interface with the MLPs may need to be improved. In our analysis, we examined various cases where (1) the full potential of the LCACs to take aboard supplies from all the MPF(F) ships is feasible and (2) certain types of supplies must be flown ashore from T-AKEs. The former cases are addressed in

---

6 *Lewis and Clark*–class T-AKEs can hangar two transport helicopters.

7 General Dynamics, "*Lewis and Clark* (T-AKE 1) Class Dry Cargo/Ammunition Fact Sheet," January 2007. A heavy underway replenishment (UNREP) system is now in development for use by MPF(F) and future aircraft carriers. An UNREP system capable of moving loads up to 12,000 pounds has been demonstrated.

the main body of this report. The latter cases are addressed in Appendix A, which also discusses the implications of requirements that certain types of supplies must be flown ashore from T-AKEs (essentially that operational flexibility and maximum throughput from the sea base will be reduced somewhat as some advantageous connector/payload pairings are excluded).

CHAPTER THREE

Major Combat Operations

The capabilities provided by reduced MPF(F) assets in MCO are explored here. Capabilities are considered in the contexts of sustaining a single MEB or two MEBs. Specifically, our analysis considered (1) the implications of eliminating one or more MPF(F) large flight deck ships (the two LHA[R]s and the LHD) and (2) the possibility of using only MLPs for sustainment. The analysis begins with the base case, proceeds through the elimination of MPF(F) LHA(R)s, and ends with the elimination of the legacy LHD.

As originally conceived, the MPF(F) squadron was to provide the capabilities to conduct a forcible entry operation, with a single MEB placed ashore within a single cycle of darkness, and provide long-term sustainment for the MEB. The MPF(F) squadron was also expected to provide casualty evacuation and to treat the MEB's casualties. The USMC no longer views the MPF(F) squadron as usable in conducting forcible entry operations. Accordingly, the MCO analysis discusses capabilities for sustainment, casualty evacuation (CASEVAC), and medical treatment.

Sustaining a Single MEB in an MCO

We begin by describing sustainment capabilities provided by components of the MPF(F) in the context of supporting a single MEB in heavy combat in an MCO. We then demonstrate means to increase those capabilities by substituting CH-53K helicopters for MV-22 aircraft. Finally, for single-MEB sustainment, we examine residual capabilities resulting from eliminating one or more flight deck ships from the MPF(F).

This baseline case uses all planned MPF(F) assets to sustain a single planned 2015 MEB in an MCO. In RAND's 2007 sea basing study, we learned that the Marines planned to conduct such sustainment operations at a distance of up to 110 NM from the MEB using only aerial connectors (CH-53K and MV-22 aircraft), and our analysis indicated that such sustainment was feasible.[1]

Since our 2007 study, the daily tonnage requirements of the planned 2015 MEB have increased, mostly due to higher projected fuel and ammunition consumption. This tonnage increase decreases the distance from ship to MEB at which air-only resupply is feasible. Material provided by the Marine Corps Combat Development Command (MCCDC) for this study indicates that the Marine Corps now intends to include surface connectors (i.e., LCACs) from the MPF(F) LHD in such sustainment operations, where some portion of the MEB's supplies will be delivered to the beach or small ports and then moved inland by truck or aircraft.

The Joint Seabasing Logistics Model (JSLM), a RAND-developed seabasing simulation, was used to model MEB sustainment with and without LCACs, using different mixes of air connectors, and with a reduced number of LHA(R)/LHD large decks. JSLM and MCCDC analysts categorize sustainment requirements identically. As a result, JSLM can use inputs from MCCDC's Capability Development Document (CDD) analysis for MPF(F) and does so.[2] JSLM simulates the

---

[1] Button et al., 2007.

[2] Sustainment requirements include ammunition; dry stores; bulk petroleum, oil, and lubricants (POL); and bulk water. Both analyses used the elements of the 2015 MEB Air Combat Element: 48 MV-22 and 20 CH-53K aircraft plus six UAVs. The operational availability of MV-22 aircraft was taken to be 82 percent, with five operationally available MV-22 aircraft withheld for casualty evacuation and other missions (for a total of 34 MV-22 aircraft used in sustainment). Operational availability the CH-53K was taken to be 80 percent. With no CH-53K aircraft withheld for other missions, a total of 16 CH-53K aircraft are used in sustainment. The operational availability of LCACs having undergone a service life extension program was taken to be 95 percent. With 18 LCACs on three MLPs, 17 LCACs are used in sustainment. This matter requires some additional discussion. The historic rate at which LCACs lose operational availability has been about 6 percent per day. However, as

operation of sustainment assets at full capacity for indefinite sustainment (i.e., at a pace that can be maintained for a considerable period of time, as opposed to surge operations that can be maintained for only a few days). For presentation purposes, JSLM consolidates sustainment and lift requirement outputs using the simple metric of aggregate short tons per day.

Sustainment was considered for distances of 25 to 110 NM from the LHA(R)/LHDs to the MEB; the results are shown in Figure 3.1. The lower curve shows that using only CH-53K and MV-22 aircraft, approximately 2,500 tons of throughput per day is achievable when the LHA(R)/LHDs are 25 NM from the MEB. With 967 tons of required daily throughput required to sustain the MEB, a capacity of 2,500 tons per day is more than twice that needed for sustainment. As the distance from the LHA(R)/LHDs to the MEB increases, flights take longer and

Figure 3.1
Tons per Day Requirement and Lift Capacities in MEB Sustainment

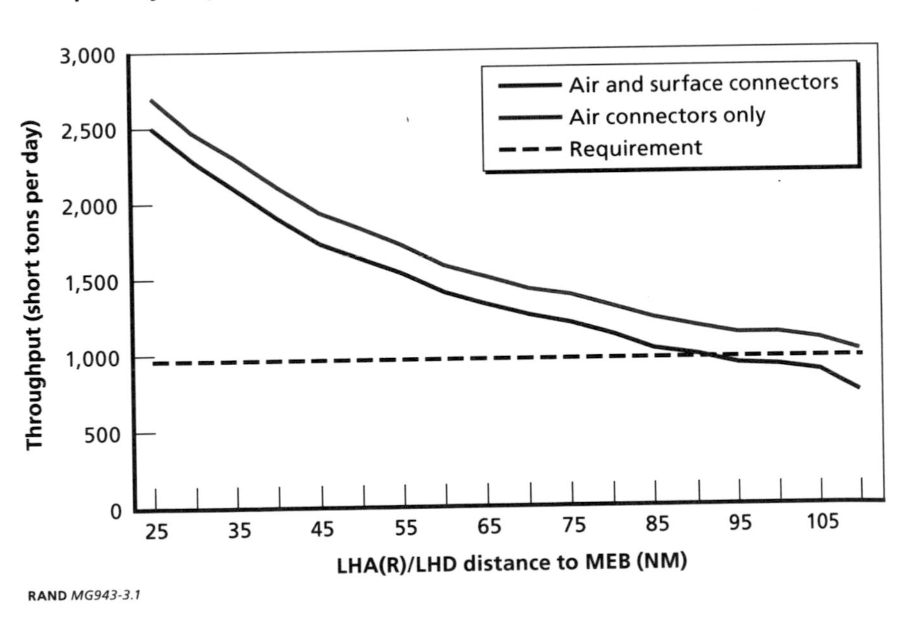

RAND *MG943-3.1*

---

future LCACs are expected to be more reliable than existing LCACs and (as in the CDD analysis for the MPF[F]), it was assumed that 95-percent availability could be maintained.

the resulting additional fuel requirements erode aircraft useful payload capacity. Consequently, as this distance increases, fewer sustainment sorties can be flown each day, and each sortie can carry less sustainment payload. Using the new MEB daily tonnage requirements, aerial sustainment of the 2015 MPF(F) MEB from a distance of 110 NM is not feasible. With the addition of the three LCACs carried by the squadron's LHD, sustainment of the MEB at 110 NM is possible with some difficulty, assuming that trucks or other means are available to move supplies inland from where the LCACs drop them at the beach.

This finding points to the need for LCACs in MEB sustainment. It also indicates the high level of consistency between results generated using JSLM and sea basing analyses from MCCDC.[3]

The metric of relative lift capacity (defined as the ratio of maximum sustainment capacity in tons per day) to average sustainment requirement (also in tons per day) offers another way to interpret sustainment results. This metric can be viewed in several ways.

Relative lift capacity reflects the robustness of available lift resources. For example, this and other studies on the issue assume favorable operating conditions but recognize that high sea states and other factors can degrade sustainment performance.[4] As noted earlier, high sea states hinder ship-to-ship transfer, slowing LCACs and reducing their capacity.[5] Potential loss of aircraft is another unfavorable factor. Recognizing the possibility of degraded sustainment capacity, a sustainment force that can barely provide a required level of sustainment under favorable conditions offers no hedge against operational degradation. High relative lift capacity, exploited under favorable

---

[3] Recall that despite the Marine Corps' preference for air-only sustainment, current planning calls for using the three LCACs aboard the squadron's LHD.

[4] For example, see "Capability Development Document of the Maritime Prepositioning Force (Future) [MPF(F)] Squadron: Increment One Mobile Landing Platforms (MLP) and Auxiliary Dry Cargo/Ammunition Ships (T-AKE)," November 16, 2007.

[5] Ship-to-ship transfer capability at the sea base is stated in terms of significant wave height. When all wave heights are measured (peak to trough), the significant wave height is defined as the mean value of the highest one-third waves. Ship-to-ship transfer is considered undegraded for significant wave heights of no more than 3 feet, or NATO Sea State 3.

conditions, can offset operational degradation experienced under less favorable conditions.

Relative capacity also reflects sustainment force flexibility under favorable conditions. A sustainment force that can provide more than the required level of sustainment can provide spare assets (such as MV-22 aircraft) that ground elements can employ to conduct tactical maneuver. Similarly, such a sustainment force can meet sustainment requirements despite aircraft losses.

Analytically speaking, high relative-lift capacity is a hedge against uncertainty; this analysis addresses the uncertain performance of future platforms, such as MLPs and the CH-53K helicopter. LCACs will undergo service life extension programs before all MPF(F) ships enter service and will be replaced in the period of interest, making future LCAC operating characteristics uncertain.[6] Recognizing these and other uncertainties, high relative-lift capacity provides a margin for error in performance estimates.

Finally, this metric can help identify and compare factors useful in achieving robust sustainment capability. For example, lift capacity metrics, such as tons per day, do not illuminate the benefits of reducing lift demand (as in counterinsurgency operations relative to heavy combat in MCOs).

Again, the relative capacity metric is the maximum throughput capacity (in tons per day) divided by the sustainment requirement (also in tons per day). The results shown in Figure 3.1 are presented in Figure 3.2 in terms of relative capacity. As above, sustainment using only air connectors is feasible for distances up to about 90 NM (at which point capacity drops below 100 percent). With the addition of three LCACs from the LHD, sustainment is feasible at distances up to 110 NM.

---

[6] A MCCDC Mission Area Analysis Branch of surface assault connectors completed in April 2006 considered possible characteristics for an LCAC replacement. LCACs having undergone service life extension are assumed here to have a maximum load capacity of 72 tons, to have a deck space of 1,809 square feet, and to average 35 knots in operation. This is consistent with the MCCDC analyses. The 2005 Naval Research Advisory Committee study of sea basing notes that LCAC speed and range are strongly affected by sea state (Naval Research Advisory Committee, *Sea Basing*, NRAC 05-2, March 2005).

Figure 3.2
Single-MEB Sustainment Using All LHA(R)/LHD Assets

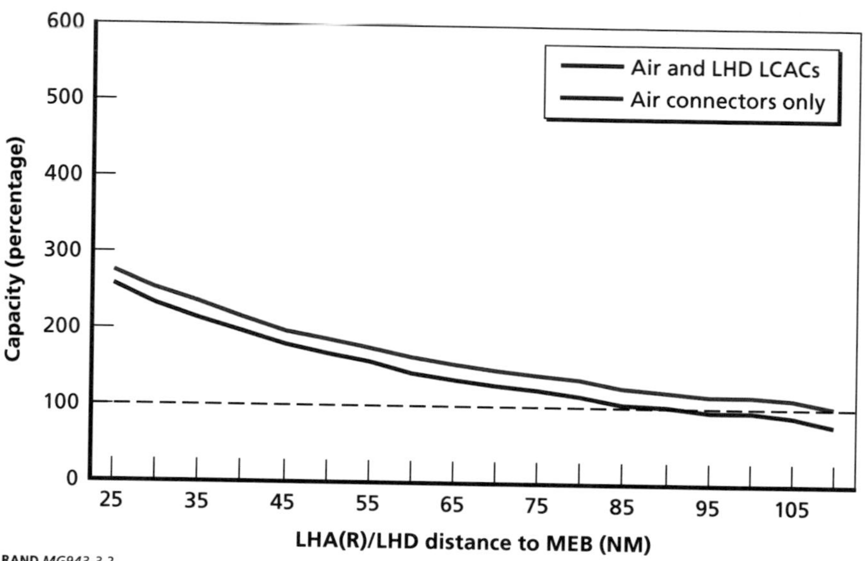

RAND *MG943-3.2*

These results do not include sustainment contributions from MLPs, which are expected to collectively provide 17 operational LCACs. We assume here that MLPs will operate 25 NM from the LCAC landing area (and that the LHD will launch its LCACs similarly). The result of adding these 17 LCACs is substantial, as shown in Figure 3.3. Lift capacity using all MPF(F) air and surface connectors is considered to be the baseline for the rest of this analysis.

We now address the results of changing the mix of air connectors, with CH-53K helicopters replacing MV-22 tilt rotor air connectors on an equal, spot-factor basis.[7] In this analysis, MV-22 aircraft reserved

---

[7] *Spot factor* is a method of sizing aircraft using the CH-46E helicopter as a unit of measure; the CH-46E is defined as having a spot factor of 1.0. The larger MV-22 has a spot factor of 2.22, and the still-larger CH-53E has a spot factor of 2.68. To illustrate, 12 MV-22 aircraft (carried on one of the LHA[R] large decks) collectively have a spot factor of 26.6 and so occupy a little less space than ten CH-53E helicopters (with collective spot factor 26.8). Conservatively, nine CH-53E helicopters can replace 12 MV-22 aircraft. The CH-53K is assumed to have the same spot factor as the CH-53E, so that nine CH-53K helicopters replace 12 MV-22 aircraft.

Figure 3.3
Single-MEB Sustainment Using All MPF(F) Connectors

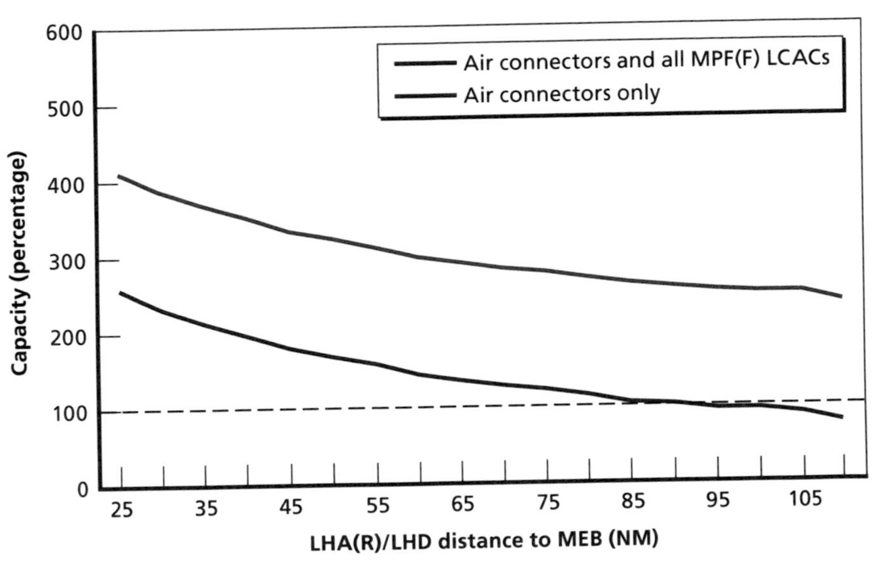

RAND *MG943-3.3*

for casualty evacuation (CASEVAC) are retained for that mission. With this mix of aircraft, 43 CH-53K helicopters are available across the three large decks instead of 16 CH-53K helicopters and 34 MV-22 aircraft. Results are shown in Figure 3.4. Again, lift capacity using all MPF(F) connectors is the baseline capacity. As mentioned, this capacity (all aircraft and up to 21 LCACs) is based on the assumption that the ships in the squadron will be capable of interacting with the key connectors—the MV-22s, CH-53Ks, and LCACs. If all the ships in the MPF(F) have the ability to offload cargo and equipment onto the LCACs, the potential throughput of the MPF(F) will be prodigious (as will be demonstrated in the following series of figures). We will also demonstrate the effects of potential constraints on throughput ability (if, for example, the T-AKEs have to be offloaded with aircraft only).

The finding that lift capacities roughly three or more times those required are achievable suggests that adequate throughput capacity could be retained with the elimination of one or more large-deck MPF(F) ships. We begin exploring the residual capability using

Figure 3.4
Single-MEB Sustainment With CH-53K Helicopters Replacing MV-22 Aircraft

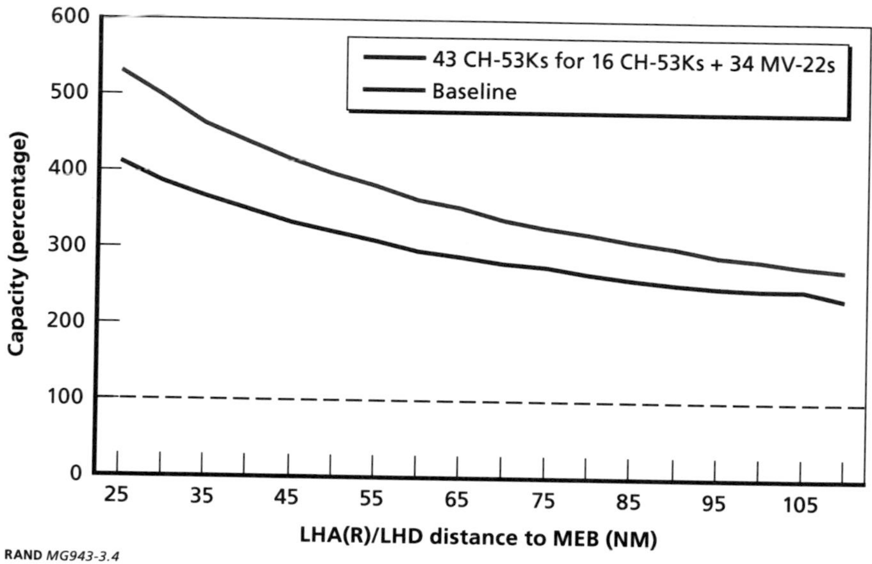

RAND *MG943-3.4*

our original mix of connector aircraft; the result of eliminating one LHA(R) from the planned MPF(F) squadron is shown in Figure 3.5. As would be expected, the loss of a single LHA(R) significantly reduces throughput capacity from the baseline, although by less than a third; LCAC capacity from the LHD and the three MLPs is unaffected (with LCACs deployed 25 NM from their landing site).

With two to four times the required throughput achievable following the elimination of one LHA(R), this scenario provides a robust residual throughput capability. We note, however, the loss of casualty evacuation and medical capabilities that were provided by the eliminated LHA(R).[8]

We now return to the option of changing the mix of air connectors, with CH-53K helicopters replacing MV-22 tilt rotor air connec-

---

8 Medical facilities in the MPF(F) squadron are situated almost entirely on the LHD and the two LHA(R)s. The LHD has six operating rooms capable of providing resuscitative surgery and 60 beds. Each LHA(R) has two operating rooms and 24 beds. All other MPF(F) squadron ships have basic sick-call facilities.

Figure 3.5
Single-MEB Sustainment Without One LHA(R)

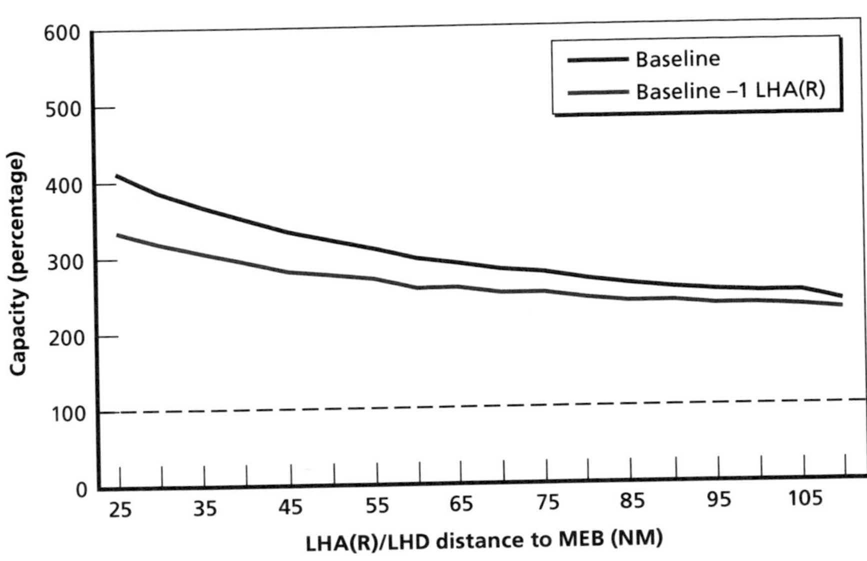

RAND *MG943-3.5*

tors on an equal spot-factor basis on the remaining large-deck MPF(F) ships. As before, remaining MV-22 aircraft reserved for CASEVAC are retained for that mission. Here, 29 CH-53K helicopters are available across the three large decks instead of 16 CH-53K helicopters and 34 MV-22 aircraft. Results are shown in Figure 3.6, with the finding that the throughput lost with the elimination of one LHA(R) is reconstituted when CH-53K helicopters are used to replace MV-22 aircraft.[9]

What if both LHA(R)s were eliminated? Figure 3.7 shows the result of eliminating both LHA(R)s with the planned aircraft mix (in blue) and with CH-53K helicopters replacing MV-22 aircraft on the

[9] We note that MCCDC throughput analyses used the concept of MV-22 load equivalents. In these analyses, CH-53K loads are roughly equivalent to three MV-22 loads. Applying this rule, 29 CH-53K helicopters are roughly equivalent to 87 (3 × 29) MV-22 aircraft in throughput capacity. Similarly, 16 CH-53K helicopters and 34 MV-22 aircraft are equivalent to 82 (3 × 16 + 34) MV-22 aircraft in throughput capacity. This is consistent with the finding that replacing MV-22 aircraft with CH-53K helicopters can offset the loss of one LHA(R).

Figure 3.6
Single-MEB Sustainment Without One LHA(R) and CH-53K Helicopters Replacing MV-22 Aircraft

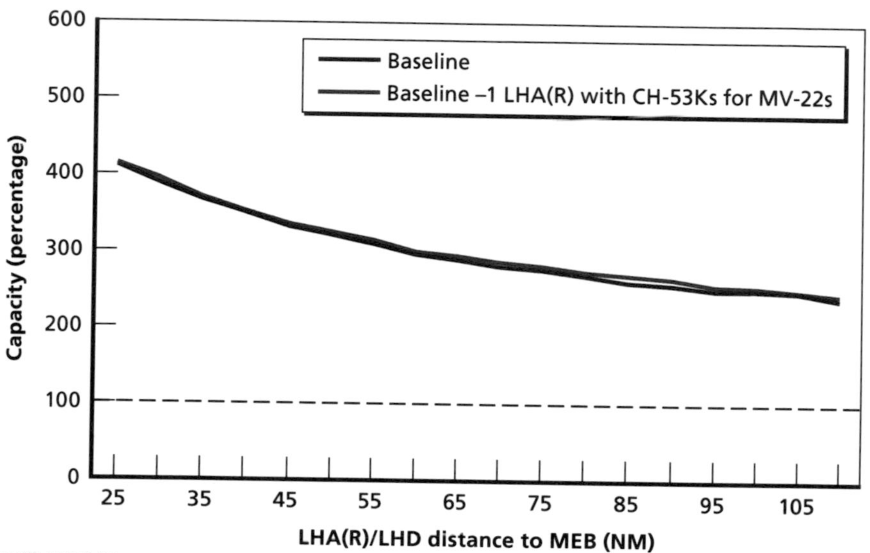

RAND *MG943-3.6*

LHD (in red).[10] A robust throughput capacity remains using air connectors from the LHD and LCACs from the LHD and the MLPs. The similarity of the results to the baseline at distances of about 100 NM suggests that with a single large flight deck, the limited number of aircraft available for sustainment makes their mix less relevant. It also suggests that at these distances, throughput from the 20 operational LCACs from the LHD and the three MLPs dominate total sustainment.

We turn now to the case of sustaining a single MEB using only LCACs from MLPs. In this instance, the only variable of interest is the distance from the MLPs to the LCAC landing area. This distance was previously fixed at 25 NM. The result of varying the distance to the landing area from 25 to 50 NM, with three or four MLPs, is shown in Figure 3.8. The sustainment capacity using three MLPs is margin-

---

[10] Throughout this study, CH-53K helicopters do not replace MV-22 aircraft reserved for CASEVAC.

Figure 3.7
Single-MEB Sustainment Without Both LHA(R)s

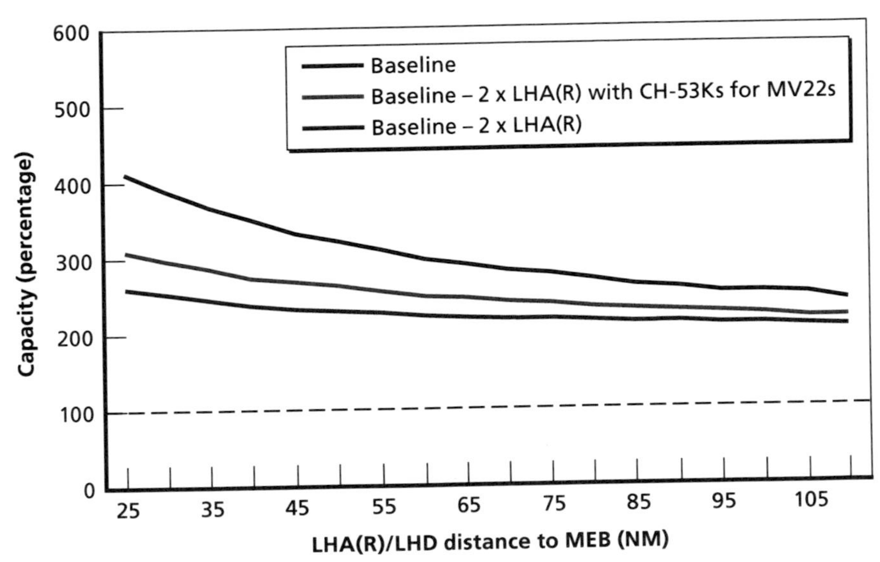

RAND *MG943-3.7*

ally adequate, with four MLPs providing a more robust capability.[11] In cases where sustainment uses only LCACs from MLPs, virtually all medical support is lost, there are no aircraft for casualty evacuation, and the MEB is completely dependent upon trucks for sustainment. As was highlighted earlier, this case is based on the assumption that cargo can be offloaded by LMSRs and T-AKEs either by transloading the cargo onto MLPs and then into the LCACs (or perhaps directly onto LCACs) for movement ashore.

---

[11] More precisely, sustainment capacity is adequate if there is no sea state degradation. As discussed in Chapter Two, sea state can be expected to be a factor up to 30 percent of the time in littoral regions of interest. This requires throughput 150 percent of the baseline capacity to make up for lost days.

Figure 3.8
Single-MEB Sustainment Using LCACs from Three or Four MLPs

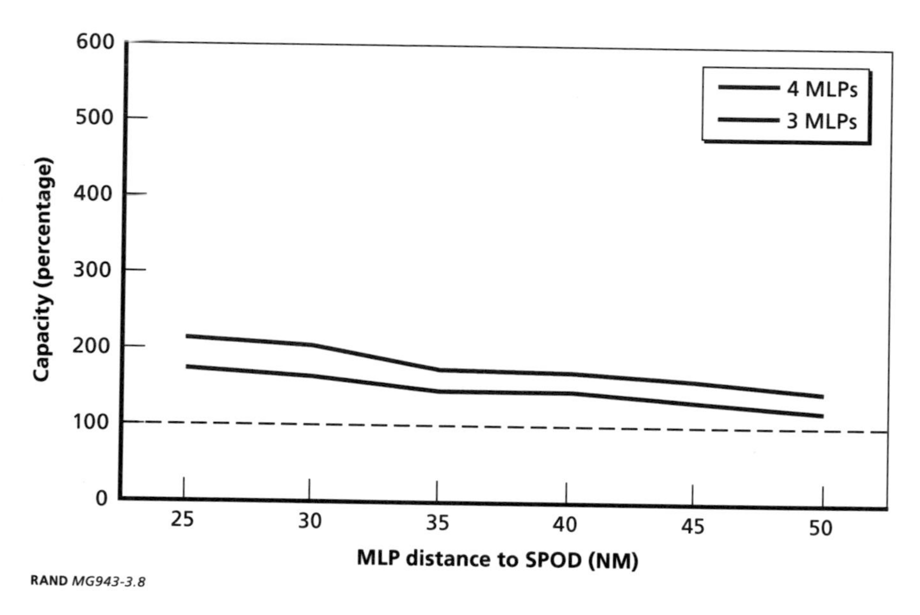

RAND *MG943-3.8*

Sustaining Two MEBs in an MCO

As described in the Chapter Two, two MEBs are sustained in MCO scenarios using the pooled resources of five ESGs and the MPF(F) squadron. Collectively, eight large-deck ships (five from the ESGs and three from the MPF[F] squadron) plus three MLPs from the MPF(F) squadron, along with their aircraft and LCACs, are employed in this effort. The mix of CH-53K and MV-22 aircraft on the ESGs' large decks is identical to that on the MPF(F) ships (described previously). The ratio of LHA(R) ships to LHDs from the 30-year shipbuilding program suggests that the five ESGs will collectively provide three LHDs and two LHA(R)s. The task of scheduling and coordinating flight and LCAC operations across eight LHA(R)/LHDs is not simple; flight and LCAC operations sometimes cannot be conducted simultaneously.[12] Further,

---

[12] We conservatively assume throughout this analysis that LHDs do not simultaneously operate their aircraft and LCACs. This reduces their daily LCAC operating time by 25 percent.

asset availability needs to be uniform over each day. If assets, such as LCACs, are clustered in time, then there will be periods in which there are more assets than needed (meaning that they will be assigned "make work" loads in order to maximize throughput), and assets will be unavailable at other times (meaning that loads cannot always be assigned efficiently). The problem can be simplified conceptually by assuming that attack aircraft (such as Joint Strike Fighters) will operate off the ESG's LHA(R)s. Under this assumption, four LHDs and two LHA(R)s (from the MPF[F] squadron) are used in the baseline case for sustaining two MEBs. Also in the baseline case, 32 CH-53K and 68 MV-22 aircraft are available for sustainment operations.[13] The four LHDs are expected to provide 12 LCACs in addition to the 17 LCACs provided by the three MLPs.[14]

Results for the case in which both air and surface connectors sustain the nearer MEB but the farther MEB is sustained by air only are shown in Figure 3.9. While the nearer MEB can be sustained comfortably using LCACs and air connectors, air-only sustainment of the farther MEB (using the planned aircraft mix) is possible out to a distance of about 70 NM.[15] The result of replacing MV-22 aircraft with CH-53K helicopters is shown in Appendix A.

When both aircraft and LCACs can sustain both MEBs, a robust level of sustainment is possible at all distances considered (Figure 3.10). The penalty for this capability is, of course, the dependence on trucks and/or aircraft, along with a reasonably secure area for truck and/or aircraft movement.

Lift capacities under this scenario are lower than was seen in single-MEB sustainment but are still substantial. This suggests that adequate throughput capacity could be retained with the elimination of one or more large-deck MPF(F) ships. As before, we now begin

---

[13] This is twice the number of CH-53K and MV-22 aircraft assigned to the MPF(F) squadron.

[14] The CDD analysis for MPF(F) uses 95-percent availability (rounded down). With this rounding and 30 total LCACs in operation, we expect a single LCAC to be inoperable at any time.

[15] This figure ignores throughput degradation through periodic high sea states.

Figure 3.9
Air-Only Sustainment of Further MEB Not Feasible

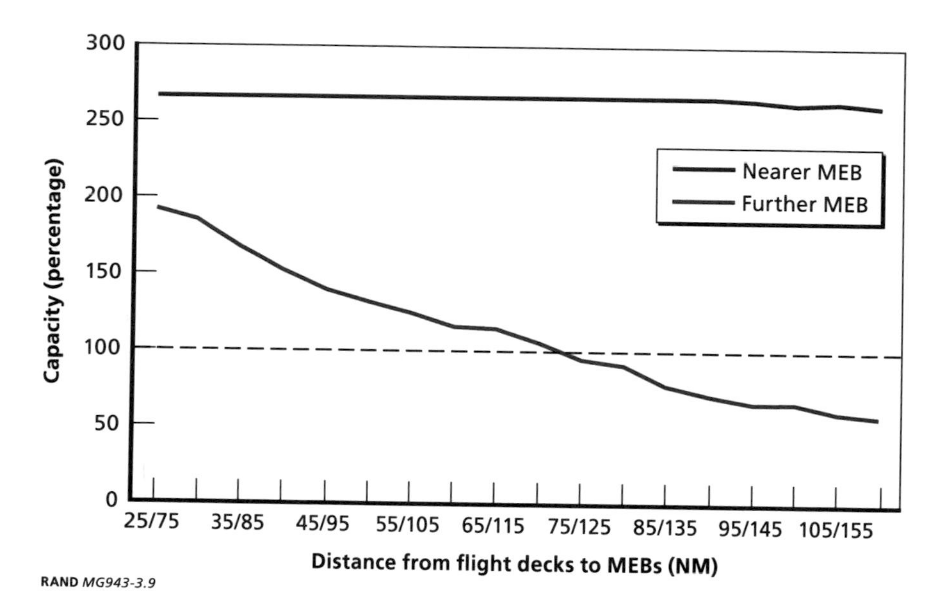

RAND *MG943-3.9*

Figure 3.10
Two-MEB Sustainment by Surface and Air Possible

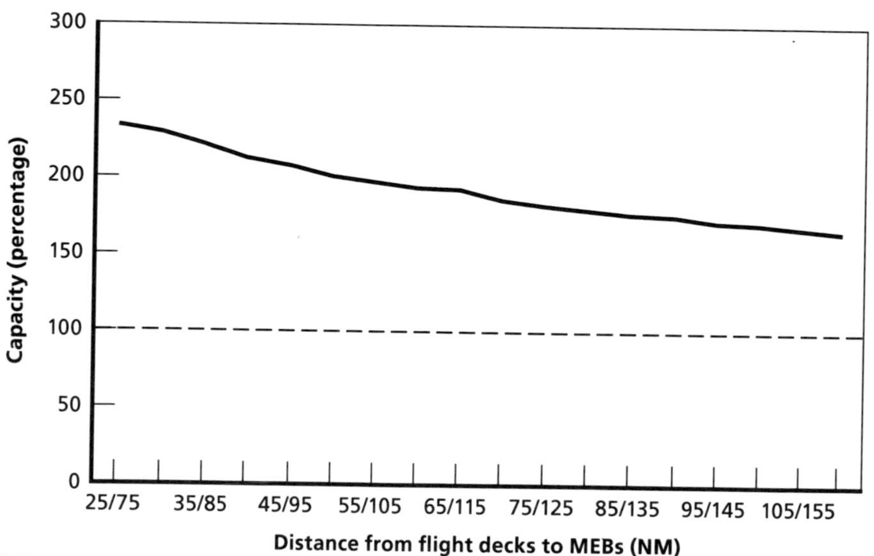

RAND *MG943-3.10*

exploring the residual capability using the original mix of connector aircraft; the result of eliminating one LHA(R) from the planned MPF(F) squadron is shown in Figure 3.11. Also as before, the loss of a single LHA(R) significantly reduces throughput capacity from the baseline level, but capacity without one of the LHA(R)s remains over 150 percent. The substitution of CH-53K helicopters for MV-22 aircraft (as shown in Figure 3.12) offsets the removal of one LHA(R) under these conditions.

As in the case of our single-MEB sustainment analysis, we now address the consequences of eliminating a second LHA(R). Figure 3.13 shows the result of eliminating both LHA(R)s but retaining the planned aircraft mix (in blue) and replacing MV-22 aircraft on the LHD with CH-53K helicopters (in red).[16] There are too few (15) CH-53Ks in the latter case to make up the loss of the LHA(R)s. However, with ESG assets untouched, the residual throughput capacity here is adequate.

Figure 3.11
Two-MEB Sustainment Without One MPF(F) LHA(R) Possible

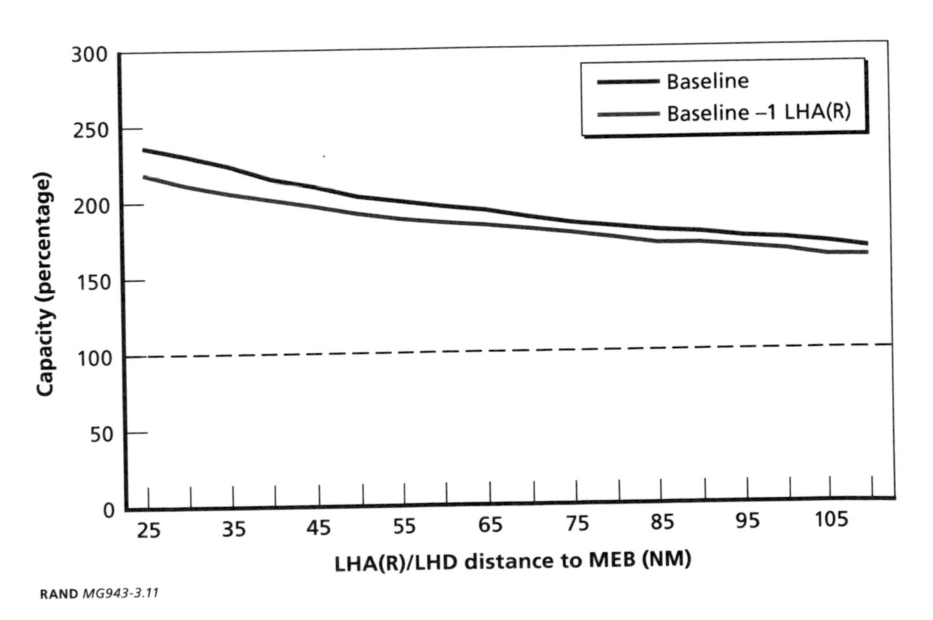

RAND *MG943-3.11*

---

16 As before, CH-53K helicopters do not replace MV-22 aircraft reserved for CASEVAC.

Figure 3.12
Two-MEB Sustainment Without One LHA(R) and CH-53K Helicopters Replacing MV-22 Aircraft

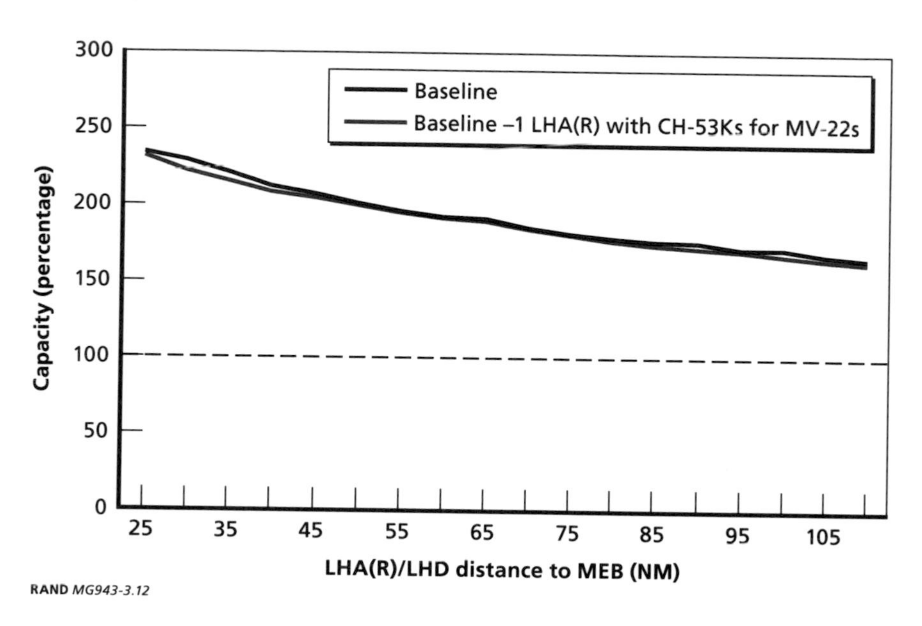

RAND *MG943-3.12*

Finally, we address the results of eliminating all large decks from the MPF(F) (with the possibility of adding a fourth MLP). Throughput capacity is adequate with three MLPs and is robust with four MLPs (see Figure 3.14). Here, the presence of the Amphibious Task Force (ATF) dilutes the consequences of removing the large flight decks from the MPF(F). This is true both in terms of sustainment throughput and in terms of other considerations, such as medical support. For example, eliminating both LHA(R)s and the LHD from the MPF(F) squadron essentially stripped it of CASEVAC and medical support capabilities. However, the presence of the ATF provides capabilities for CASEVAC and medical support.

Figure 3.13
Two-MEB Sustainment Without Both MPF(F) LHA(R)s

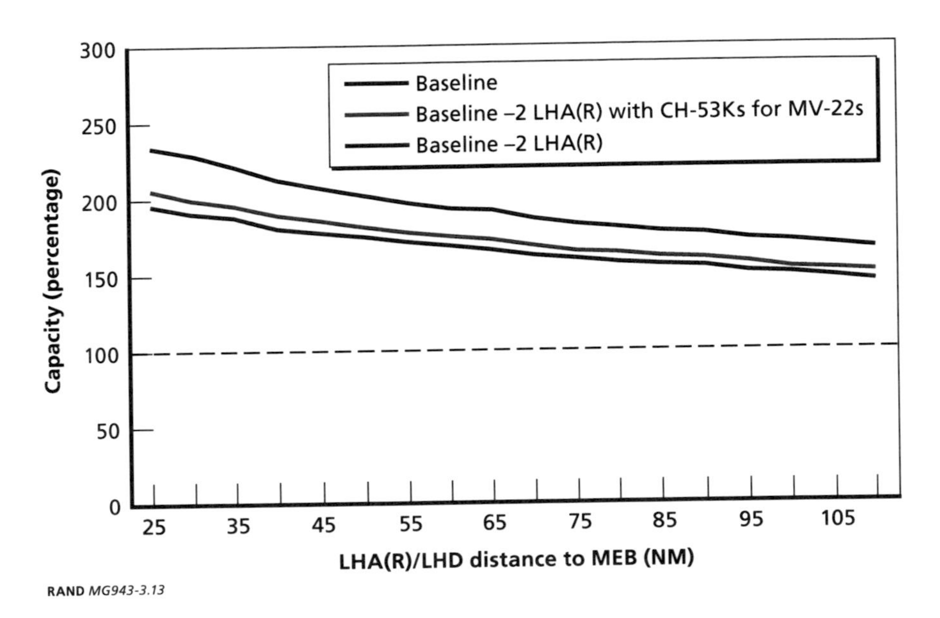

RAND *MG943-3.13*

Figure 3.14
Two-MEB Sustainment Without Both MPF(F) LHA(R)s and LHD

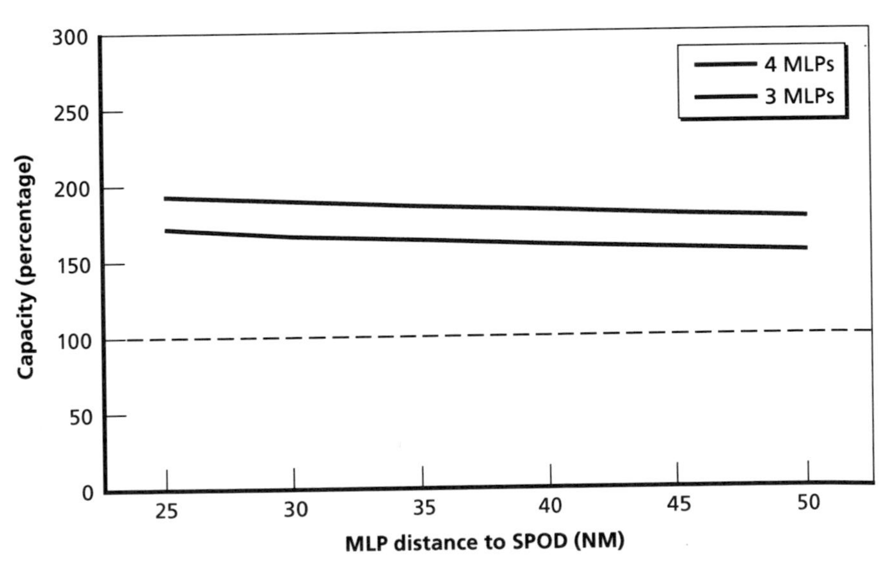

RAND *MG943-3.14*

CHAPTER FOUR

Counterinsurgency Operations

In addition to using the sea base to support ground combat forces in MCOs, there could be considerable application for a sea base in support of COIN operations. This chapter will first examine the nature of COIN operations and then discuss how the sea base, specifically the MPF(F) component, could support COIN.

The Nature of COIN Operations

There are some important differences between MCO and COIN operations:

- *Duration.* Whereas MCO tend to be of relatively short duration, COIN operations tend to be far more protracted. On average, insurgencies last approximately 12 years.[1]
- *Restricted rules of engagement.* There tends to be far less restriction on the use of lethal force in MCOs compared to COIN operations. Indeed, in COIN, the key objective is usually winning and maintaining the support of the population. Therefore, firepower must be applied in a very constrained manner to minimize the chances of collateral damage and civilian casualties. This reality tends to reduce ammunition consumption in COIN operations compared to MCOs.

[1] Walter L. Perry and John Gordon IV, *Analytic Support to Intelligence in Counterinsurgencies*, Santa Monica, Calif.: RAND Corporation, MG-682-OSD, 2008, pp. xi–xii.

- *Role of indigenous forces.* It is ultimately the local government and its security forces that will win or lose the COIN effort. While foreign military forces (U.S. or other) can be of considerable assistance to the local police, military, and intelligence organizations, ideally the foreign forces are in a supporting, enabling role as opposed to direct engagement in combat operations. This desire to place indigenous forces in a leading role could lead to the MPF(F), or portions or it, supporting foreign troops engaged in COIN.
- *Political considerations.* Because COIN is ultimately about the legitimacy and support of the population for the local government, political considerations often trump military ones in COIN operations. For example, the U.S. military units committed to a COIN operation may have to use sea basing for a protracted period because the presence ashore of large numbers of U.S. soldiers could reinforce the insurgents' propaganda.

Possible Roles of the Sea Base in COIN

Given the type of environment that U.S. forces will face in COIN, what are the possible missions that a sea base, especially the MPF(F) element, could accomplish? The primary role of the MPF(F) in an MCO is to help introduce and sustain a MEB and possibly other joint forces.[2] In a COIN operation, the United States could participate with considerable combat forces, as was the case in Iraq, Afghanistan, and Vietnam, or the number and mission of U.S. forces could be much smaller and different, with U.S. forces training and advising local forces and possibly providing key enabling capabilities for them. It is possible that all or a portion of the U.S. forces could be supported, at least initially, from a sea base.

The most valuable use of a sea base in MCOs would likely be early in the conflict, when U.S. forces might have a limited number

[2] For a detailed examination of how and to what extent the MPF(F) might be able to support an Army brigade in addition to a MEB, see Button et al., 2007.

of bases ashore in the operational area. Once the area of operations becomes more developed, U.S. forces would probably place a greater reliance on traditional ports and airfields to introduce and sustain additional forces. Indeed, the sea base will always be constrained in the total number of units that it can sustain, with two to three brigade-size units probably being the upper limit that the MPF(F) squadron can support. The role of the sea base later in MCOs might be to supplement the capacity of traditional ports in the event they suffer further damage from long-range precision strikes or to serve as the base from where forces might be introduced into a previously unoccupied portion of the area of operations. Importantly, the expectation would probably be that intense MCOs would be wrapped up in a relatively short period of time, possibly a few days or weeks.

In COIN, however, the duration of the operation will almost certainly be far longer—many months or years. What would the role of the sea base be in those circumstances? As mentioned above, U.S. forces engaged in COIN could perform direct combat missions, mostly engage in assisting and enabling local forces, or undertake some combination of the two. When U.S units are engaged in combat operations, the sea base could provide the means of entry and the initial sustainment for U.S. forces. In that situation, the role of the MPF(F) would be generally similar to the mission that is envisioned for it in MCOs, although depending on the size of the U.S. force requiring support from the sea base, it might not be necessary to deploy an entire MPF(F) squadron; a portion of the squadron might be sufficient.

If the political and military situation permits, the U.S. forces engaged in COIN would probably revert to traditional sources of supply—ports and airfields—for protracted operations. In those cases, it may be possible to withdraw the MPF(F) elements that were initially supporting the forces engaged in COIN. How long the presence of the sea base would be required is highly situationally dependent. It could be needed for a few days, for several weeks, or for several months. In more protracted deployments, issues will arise, such as replenishment of the ships in the sea base and rotation of their crews.

The sea base could also prove a valuable asset when U.S. forces are performing primarily a train, assist, advise, and enable role in COIN.

Here it would not be serving primarily as a base for fighting units. In situations where there are political sensitivities associated with a U.S. presence ashore, the sea base could allow U.S. forces operating alongside the forces of the host nation to minimize their "footprint" ashore. Put another way, the sea base could be the base of operations for U.S. advisors, trainers, small numbers of manned aircraft, and reconnaissance and surveillance platforms, such as unmanned aerial vehicles, as well as a hub for supplies, maintenance, and medical activities. This would allow U.S. commanders to operate from the sea base, sending ashore only those personnel and equipment necessary for the mission.

When used to provide a base for U.S. advisors and trainers, the sea base could be home for both conventional and special operations forces. The numbers and mix of conventional and SOF personnel is, of course, highly situationally dependent. When SOF are aboard the sea base, they will probably require security and communications arrangements that are not normally available aboard the MPF(F) ships.

In addition to providing a train-and-assist role for the indigenous forces, U.S. units on the sea base could also be providing key enabling capabilities that the local police, intelligence, and military units do not possess. For example, the local forces may have either no or few unmanned aerial vehicles (UAVs) for reconnaissance and surveillance use. The MPF(F) could then act as the base for U.S. UAVs that would be flown in support of the local forces, with the data from the UAVs being made available on the ships, as well as to U.S. advisors who are embedded with the local forces.

Finally, the sea base could be used to transport and supply the forces of the host nation. There are a number of littoral nations (such as the Philippines or Indonesia) that might benefit greatly from this capability. Here, the sea base could be used to transport elements of the local military to a new operational area and then act as the hub from which those forces are supplied and provided other forms of support (such as the UAV example mentioned above or CASEVAC back to the ships where their injured personnel would receive initial medical care). How many MPF(F)—or other—ships might be needed to transport and supply foreign forces is clearly situationally dependent. Importantly, many of the ships in the planned MPF(F) squadron will be preloaded with the supplies and equipment for a MEB, which would require those ships be either partly or entirely unloaded in order to support foreign forces. There is, however, the possibility that a portion of the MPF(F) squadron could be deployed to assist foreign forces in COIN. For example, one or more MLPs and one or more flight deck ships (LHD or LHA[R]) could be temporarily withdrawn from Guam and deployed for the COIN mission, returning to the main body of the squadron when the situation allows.

It should be noted that if the MPF(F), or portions of it, are used to support foreign forces in what could be a protracted COIN operation, the enemy might be able to identify and try to exploit potential vulnerabilities. For example, if MPF(F) ships are operating somewhat close to shore for extended periods of time, the enemy might be able to infiltrate saboteurs onto the ships or to attack the ships with explosive-armed small boats. These threats would require appropriate force-protection measures be taken.

Logistics Requirements in COIN Operations

We now explore the capabilities of the MPF(F) to support a MEB in counterinsurgency. An important issue here is the level of supply (expressed in tons, by major type of supply, such as fuel, water, and ammunition) needed by a MEB conducting COIN operations. The Marine Corps has not yet undertaken a detailed study of this issue. Certainly the daily tonnage needs of a MEB in conducting COIN could vary greatly, and would depend on the situation.

Two different levels of supply requirements were used to bound our analysis. Our "COIN-High" tonnages assume that the MEB would need 90 percent of the fuel and 50 percent of the ammunition that it would require in an MCO. Requirements for "other stores" (miscellaneous supply items) and water were assumed to be the same as in a MCO. The "COIN-Low" case assumed that the MEB would only need 50 percent of the fuel and 10 percent of the ammunition compared to an MCO, again with water and other stores requirements being the same. These assumptions are depicted in Figure 4.1 using a

Figure 4.1
Notional MEB Supply Requirements in COIN

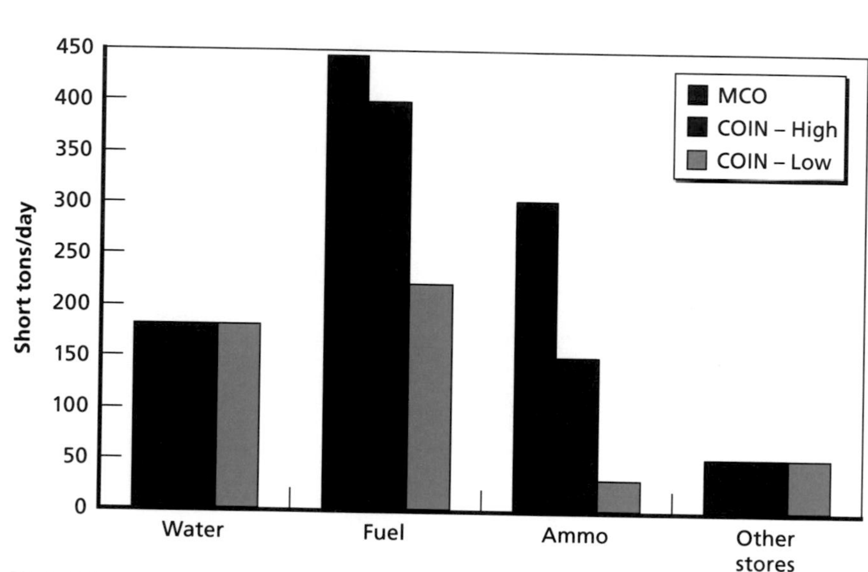

RAND *MG943-4.1*

MEB in MCO operations as a basis for comparison. The projected tonnages in Figure 4.1 are RAND-derived and not officially sanctioned by the Marine Corps or Navy.

Using the consumption factors mentioned above, various cases were examined to determine how well, or not, the MPF(F) could support a MEB engaged in COIN.

Figure 4.2, below, shows the results when the program of record (three flight decks, three MLPs, and associated supply ships) is supporting a MEB with "air only"—in other words, no LCACs were used; the only connectors are CH-53s and MV-22s (in the planned numbers).

The results depicted in Figure 4.2 indicate a robust air-only sustainment capability out to 110 NM with the lower rate of consumption (the red line). With the higher consumption rate MEB (the blue line), air-only sustainment is not quite feasible when the MPF(F) LHD and LHA(R)s operate 110 NM from the MEB.

The effect of adding just three LCACs from the MPF(F) LHD is shown in Figure 4.3. Note that at the shorter distances from the ships

Figure 4.2
Air-Only Resupply of a MEB in COIN

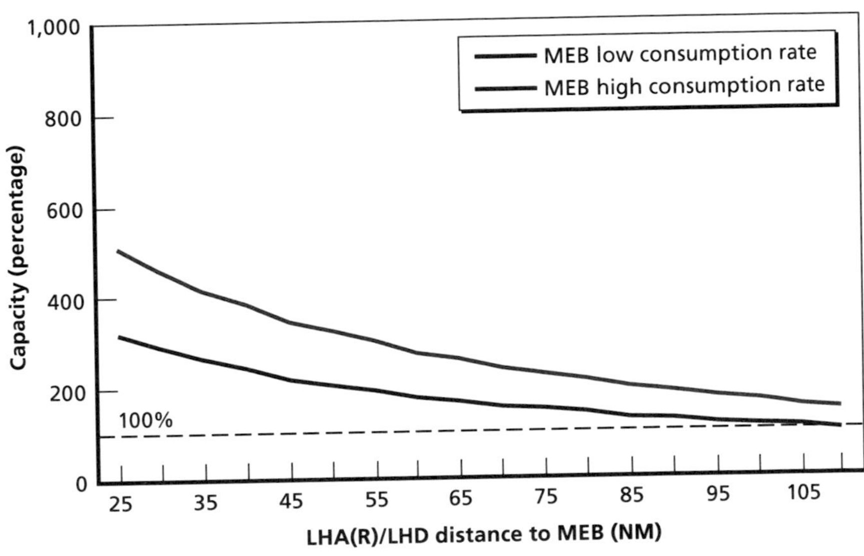

RAND *MG943-4.2*

Figure 4.3
LHD LCACs and Aircraft Used to Support MEB in COIN

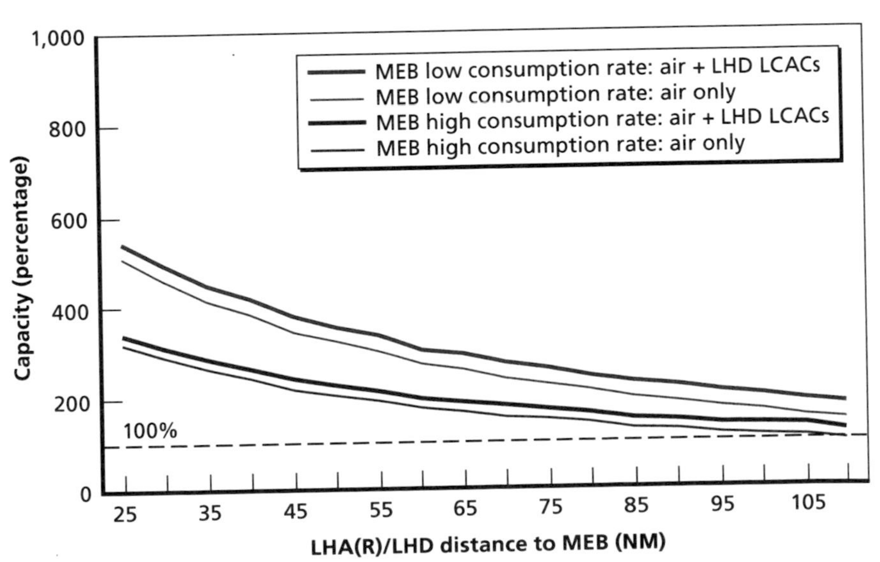

RAND *MG943-4.3*

to the MEB (roughly 25 NM or less) there is a considerable amount of overcapacity. This is particularly significant in light of the observation that the threat to MPF(F) ships supporting COIN operations can be expected to be lower than threat levels in MCOs. Consequently, it may be feasible for the MPF(F) ships to operate significantly closer to the coast in COIN operations than in MCOs.

Next, we add the 17 LCACs expected to be operational from the three MLPs to the connectors available to move supplies ashore. Here, significant overcapacities are seen, with at least three times the required throughput capacity at all ranges and at least five times the required throughput capacity for operations conducted at short distances. As was the case in the MCO analysis, the throughput potential shown in Figures 4.3 and 4.4 are based on the assumption that LCACs could load supplies from various MPF(F) ships including the T-AKEs.

These cases show the capability of the program of record MPF(F) in supporting a MEB in COIN operations, given our assumptions of high and low level of daily consumption rates. The next logical ques-

Figure 4.4
All MPF(F) Aircraft and LCACs Resupply a MEB in COIN

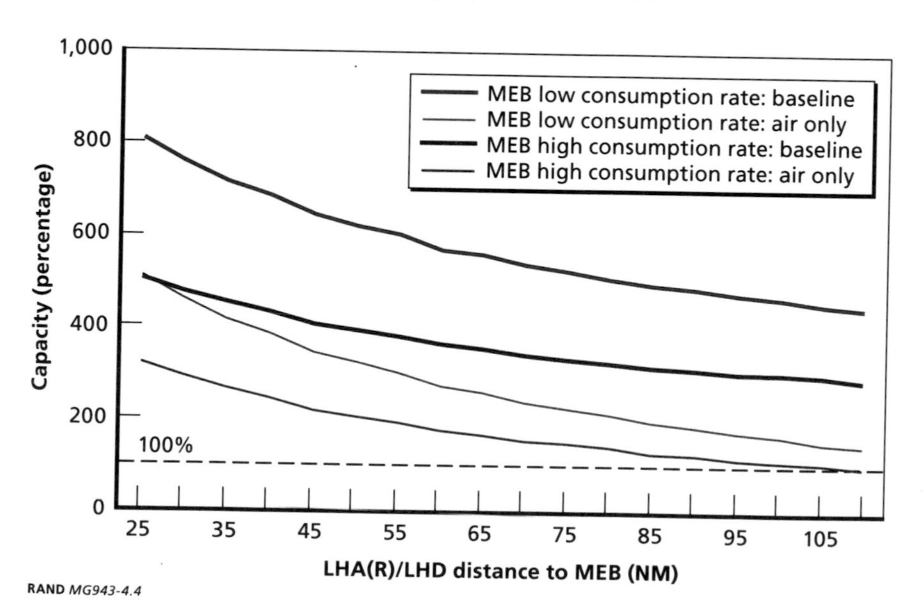

RAND *MG943-4.4*

tion was how would that change if the 14-ship MPF(F) squadron were modified—in particular, if there were fewer flight decks available?

The results of eliminating one LHA(R) and its aircraft are illustrated in Figure 4.5. The red lines depict the base case—all programmed aircraft and ships—whereas the blue lines show the result of removing one LHA(R). Note that the MEB's daily tonnage needs are met with substantial margin remaining.

As in our MCO analysis, we proceeded to alter the aircraft mix to see if substituting CH-53Ks for MV-22s could compensate for the removal of one LHA(R) and its aircraft. The results (shown in Figure 4.6) are consistent with the MCO observations—when CH-53Ks replace MV-22s, the overall tonnage capacity is restored to original levels. This reinforces results seen earlier in the case of MCOs: In a purely logistical sense, substituting CH-53K heavy lift helicopters for MV-22 aircraft can offset the deletion of one MPF(F) LHA(R).

We also examined a case where only the MLPs were available to support the MEB. In this case, there are no large flight deck ships

Figure 4.5
MEB in COIN Supported by MPF(F) Minus One LHA(R)

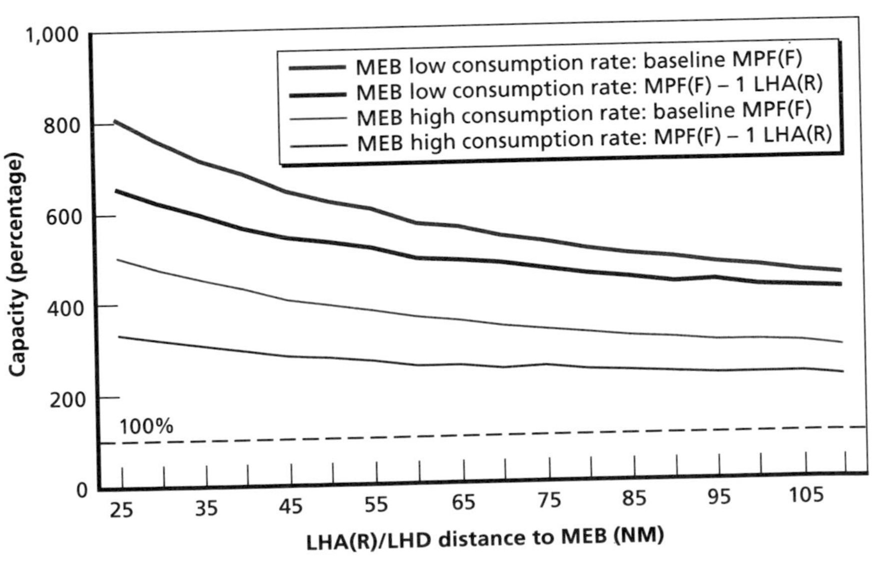

RAND *MG943-4.5*

Figure 4.6
MEB Supported by MPF(F)—Minus One LHA(R)—With CH-53Ks Substituted for MV-22s

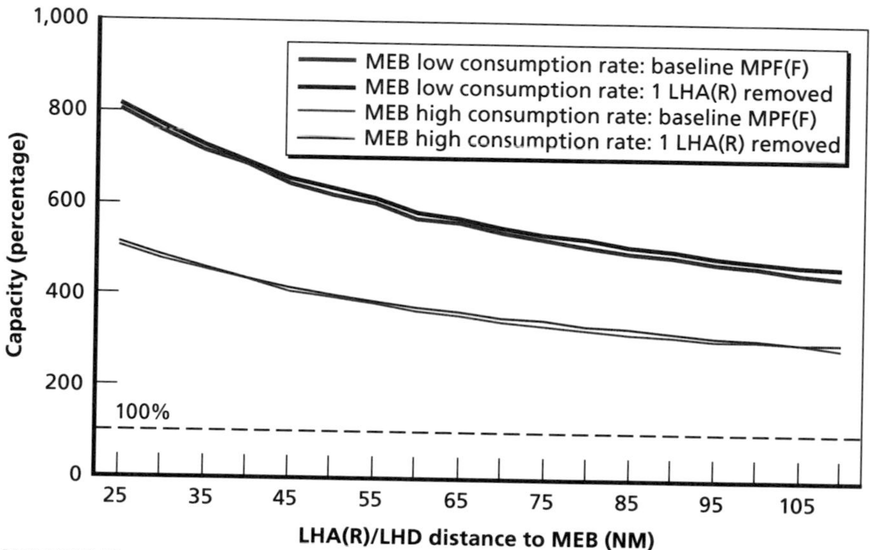

RAND *MG943-4.6*

(LHA[R] or LHD) or their aircraft—all supplies are moving from ship-to-shore via LCACs. Figure 4.7 illustrates that the MLP/LCAC combination has more than enough capacity to support the daily tonnage requirement of a MEB engaged in COIN.

However, important operational constraints could result if all the air-capable ships were removed from the MPF(F). If both LHA(R)s and the LHD were deleted, the ability of the MLP to operate with T-AKEs could be constrained by the relatively small number of transport helicopters carried for VERTREPs by T-AKEs. There would be no MV-22 tilt-rotor aircraft and only three or four total deck spots on the MLPs (one per ship) available for casualty evacuation. Additionally, the superior medical facilities of the three air-capable ships would be lost if all were removed. Finally, the loss of the aircraft could place tactical constraints on the MEB commander. So, while our analysis showed that in terms of raw throughput capacity the three MLPs and their LCACs could meet the daily tonnage requirements of the MEB,

Figure 4.7
MEB in COIN Supported by MLP LCACs Only

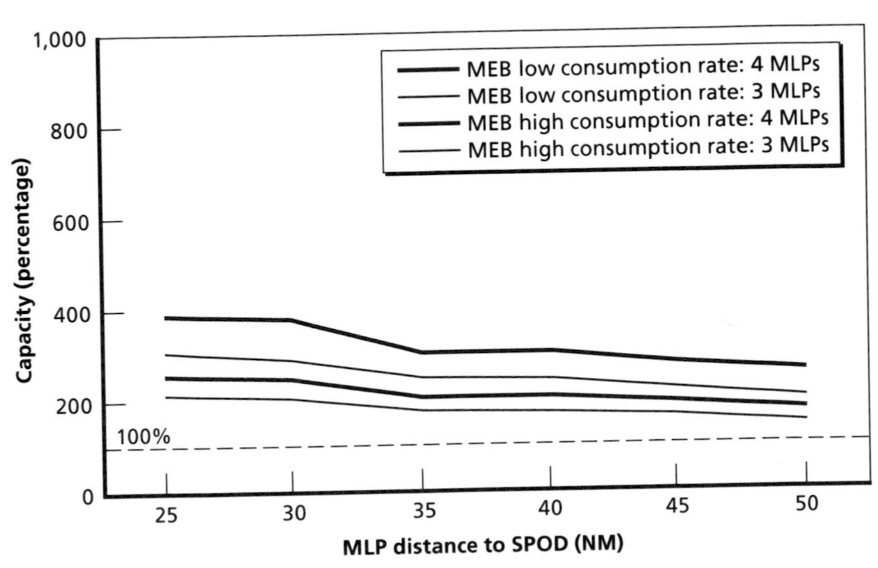

RAND *MG943-4.7*

significant operational issues could arise with no air-capable ships in the force.

CHAPTER FIVE

The MPF(F) as a Joint Special Operations Task Force Afloat Forward Staging Base

Introduction

SOF conduct a broad spectrum of military missions, from supporting MCOs with special reconnaissance to conducting small, independent foreign internal defense or civil affairs missions. In large, conventional military operations, commanders may establish a joint special operations task force (JSOTF) ashore, such as the Combined Joint Special Operations Task Force (CJSOTF) headquartered at King Fahd International Airport, Saudi Arabia during Operation DESERT SHIELD and Operation DESERT STORM in 1990–1991. However, in crisis response operations, SOF have more frequently relied on ships and sea basing to carry out operations in support of MCOs. Consider SOF support for Operation RESTORE DEMOCRACY in 1994–1995 (based at sea off the coast of Haiti) and of Operation ENDURING FREEDOM in 2001 (likewise based off the coast of Afghanistan). In both cases, commanders based SOF aboard aircraft carriers, the USS *America* and USS *Kitty Hawk* respectively.[1]

When an international crisis emerges, the ability of a combatant commander to rapidly deploy a JSOTF to a forward staging base (FSB) and begin conducting Phase I operations can be crucial to mission success. Sea basing SOF on an afloat forward staging base (AFSB) has allowed early access to theaters throughout the history of U.S. Special Operations Command (USSOCOM), from relatively small mis-

[1] U.S. Special Operations Command, *United States Special Operations Command History 1987–2007*, 2007a.

sions, such as noncombatant evacuation operations (NEOs) and limited humanitarian assistance, to larger, forced-entry operations, such as Phase II of Operation IRAQI FREEDOM. However, USSOCOM's access to U.S. Navy shipping has always been limited and ad hoc. This chapter will discuss the advantages and disadvantages of basing SOF afloat and explore the suitability of the developing MPF(F) platforms for that course of action in the future.

The Nature of Joint Special Operations

The previous chapter on COIN addressed the utility of employing the MPF(F) to support and sustain many operations that include SOF. For example, psychological operations, civil affairs, and foreign internal defense are special operations often conducted under the broader umbrella of COIN that generally have the same sea basing requirements as conventional forces. In this chapter, we will discuss special operations missions that go beyond COIN and demand unique support so that we can understand how sea basing may best support SOF. These more demanding SOF missions include

- Phase II MCO operations
- counterterrorism
- unconventional warfare.

High-risk, vital targets, elusive intelligence, and small, specialized tactical units are common characteristics of special operations set in geopolitically semi- to nonpermissive and/or medium- to high-threat environments. In order to mitigate the high risk of these missions, SOF commanders rely on a number of key principles:[2]

- *Surprise.* Unlike Phase III decisive combat operations, small units operating in high-threat environments rely heavily on preserving

[2] William H. McRaven, *Spec Ops: Case Studies in Special Operations Warfare Theory and Practice*, Novato, Calif.: Presidio Press, 1995; and U.S. Joint Chiefs of Staff, *Joint Special Operations*, Joint Publication 3-05, 2006.

the element of surprise to maintain their freedom of maneuver and effectively prepare the battlespace for Phase III.[3] Moreover, for fleeting targets in counterterrorism operations, surprise often means the difference between a dry hole and a successful capture. The loss of surprise against a barricaded, high-value target can mean the difference between a successful capture and resistance that prompts a killed target (and a lost opportunity to interrogate and learn more about a complex terrorist network).

- *Precision and speed.* Because SOF units are small, they rely on precise targeting and speed to execute their missions during brief windows of opportunity before an enemy can respond and overwhelm them with larger forces. Detailed intelligence helps SOF determine precisely where enemy and friendly combats are located, what systems are functioning, and how to most effectively accomplish their missions with the least risk. And extensive training and numerous mission rehearsals prior to launch ensure that teams can respond rapidly when a target emerges.
- *Discretion (security).* Unlike conventional forces in MCOs, where massive units gather in assembly areas near SPODs and aerial ports of debarkation, SOF units conducting Phase II operations, counterterrorism, or unconventional warfare try to stage forces on bases as inconspicuously as possible. When SOF forces are based ashore, enemy agents can easily monitor the departure of aircraft, boats, or vehicles and tip off an impending assault. Moreover, land bases expose friendly agents to enemy agents monitoring their comings and goings.

Role of Sea Bases in Special Operations

Pursuing evasive high-value targets inside partner nations worried about their image as a competent, sovereign nation; discreetly advising

[3] The U.S. military model divides the prosecution of war into four phases: Phase I, deterrence and engagement; Phase II, seize the initiative; Phase III, decisive operations; and Phase IV, post-conflict operations.

friendly agents to operate against a pariah state; secretly placing eyes on target prior to decisive combat operations: These are missions that require sophisticated planning and execution. A JSOTF conducting these missions requires several maneuver and support elements to fully support the main efforts ashore and be prepared for high-risk contingencies in often challenging, time-sensitive environments, including

- a JSOTF headquarters with an interagency intelligence center
- special reconnaissance teams
- assault teams
- quick reaction forces
- air mobility; e.g., a joint special operations air component (JSOAC) including MV-22s, MH-60s, OH-6s, AH-6s, CH-47s, and UAVs
- maritime mobility; e.g., a Naval Special Warfare Task Group (NSWTG) including rigid inflatable boats, Mk V boats, SEAL delivery vehicles, and unmanned undersea vehicles
- capabilities for
  - detainee handling
  - fire support
  - medical care
  - explosive ordnance disposal
  - intelligence, surveillance, and reconnaissance.

Capacity

MPF(F) ships have the capacity to rather easily accommodate a robust JSOTF of 1,400 personnel plus aircraft, boats, vehicles, UAVs, and supporting medical and detainee handling spaces using an MLP and an LHA(R) or LHD. With the MLP hosting the headquarters, key maneuver units, detainee handling, and medical support, the JSOAC would be free to conduct air operations off the LHA(R) or LHD without interfering with mission rehearsals or day-to-day maintenance training for the assault teams and quick reaction forces on the MLP. Collocating detainee handling, medical support, and the intelligence center ensures quick turn around of information exploited from a mission.

Security

A key advantage of sea basing in this manner is the force protection and operational security derived from a mobile platform based over the horizon away from enemy sensors, agents, or weapons. Also, by sequestering headquarters staff afloat, battle rhythm and mission focus can be maintained with fewer distractions and less risk of information leaks.

This added security is crucial for advanced special operations and unconventional warfare, when the AFSB may be supporting special reconnaissance or human intelligence operations. Elaborate supporting plans for fire support, medical evacuation, or logistics are often untenable in these high-risk missions. However, with an AFSB, commanders can provide robust options for such supporting elements, as well as improved communication connectivity, without burdening the overall campaign with a large footprint ashore.

Mobility

In terms of operational mobility and access, the AFSB can interoperate with a Joint High-Speed Vessel (JHSV) or a guided missile submarine (SSGN).[4] These platforms give a SOF commander the option of moving a reconnaissance team or assault team to a target via air, land, sea, or undersea. Teams can transfer to an SSGN and launch their mission from a Joint Multi-Mission Submersible, with a quick reaction force standing by on Mk V special boats, or on alert on the flight deck of the LHA(R) with MV-22's. With LCACs, teams can bring their own vehicles ashore at night and conduct mounted patrols to a target. The open decks and accessibility of the MLP make it especially suited to interoperate at sea with most SOF platforms and U.S. Navy fleet platforms, such as the MH-60R and MH-60S, providing command-

[4] Four *Ohio*-class SSBNs have been modified to SSGNs. In this modification, the submarine-launched ballistic missile capability has been replaced with the ability to launch up to 154 Tomahawk land-attack cruise missiles. Other new features enable SSGNs to support SOF operations by accommodating up to 66 SOF personnel, along with the facilities, equipment, and materiel needed for support of sustained SOF operations.

ers a flexible, adaptable, responsive forward staging base to deal with changing weather and enemy threats.[5]

Limitations

The capacity, security, and mobility associated with an MPF(F) AFSB can be key advantages, but there are also limitations to consider. Keeping SOF afloat for more than six months can diminish perishable skills, including marksmanship, diving, parachuting, tactical driving, and physical conditioning. Rotating forces can alleviate this limitation. The combined capacity of an LHA(R) and an MLP is suitable for a robust JSOTF, but it is limited. Also, while the sequestered nature of a sea base is excellent for discreet SOF operations, when SOF are interoperating with other conventional forces, that isolation can hinder the coordination of plans. Finally, the limited availability of suitable platforms is a key limitation that has hampered SOF sea basing in the past. The U.S. Navy does not maintain any ship specifically devoted to hosting a JSOTF afloat, so the AFSB option has been based primarily on the availability of possible platforms and not on the operational effectiveness of sea basing SOF. In the instances of Operation ENDURING FREEDOM in Afghanistan and Operation SUPPORT DEMOCRACY in Haiti, where aircraft carriers were employed in such operations, embarked units had to off load to make room for the incoming JSOTF.[6] These limitations are not difficult to surmount conceptually, but in practice they are often hindered by competing priorities. If a commander needs additional capacity, an additional platform may not be available. Weather conditions may ground helicopters, preventing planning officers from assembling in one location to coordinate plans. And MPF(F) platforms may be called away to support their primary missions in support of MCOs and not be available to support SOF.

---

[5] The MH-60R Seahawk is a medium-lift utility helicopter; its mission set includes antisubmarine warfare, antisurface warfare, antiship surveillance and targeting, VERTREP, communications relay, combat search and rescue, and SOF support. The MH-60S Knighthawk has replaced the Navy's CH-46D Sea Knight cargo helicopter. The MH-60S mission set includes combat search and rescue, special warfare support, and airborne mine countermeasure missions.

[6] U.S. Special Operations Command, 2007a.

USSOCOM's Strategic Objectives

USSOCOM clearly spells out its vision for SOF in the future in the 2006 *Capstone Concept for Special Operations.* Citing USSOCOM's unprecedented role as a supported combatant commander in the Global War on Terror and the persistent responsibilities that duty will require long into the future to effectively contain the lingering threat of terrorism, the *Capstone Concept* emphasizes the need for global expeditionary forces capable of quickly responding to emerging crises.[7]

Underpinning the concept of joint expeditionary SOF (JESOF) is the assumption that commanders will be able to negotiate FSBs in partner nations rapidly, or alternatively that U.S. Navy vessels will be available as AFSBs. This assumption is faulty for two key reasons: (1) There are no plans to make the availability of U.S. Navy vessels more than an ad hoc capability, and (2) growing anti-Americanism throughout the world makes land-based FSBs increasingly less likely. We have already discussed the limited availability of Navy platforms as AFSBs, but the issue of anti-Americanism ashore accentuates the forward basing issue even further. Since the U.S. invasion of Iraq, the Pew Research Center has found an alarming level of anti-Americanism throughout the world, and especially in predominately Muslim regions where the United States would most likely require FSBs for crisis response operations.[8] This anti-Americanism makes it more difficult for potential host-nation governments to accept JESOF on their soil in FSBs; and if they do, the agreements often come with politically driven constraints that significantly limit the effectiveness of a JSOTF. For example, caps on the number of personnel allowed in country have limited operations in the Philippines for JTF-515, and attack aircraft are forbidden

---

7 U.S. Special Operations Command, *Capstone Concept for Special Operations*, MacDill Air Force Base, Fla., 2006. See also U.S. Special Operations Command, *USSOCOM Posture Statement 2007*, MacDill Air Force Base, Fla., 2007b.

8 Andrew Kohut, "America's Image in the World: Findings from the Pew Global Attitudes Project," testimony before the Committee on Foreign Affairs, U.S. House of Representatives, March 14, 2007.

at Camp Lemonier in Djibouti, which substantially limits CJTF-Horn of Africa's operations.[9]

Conclusions on the Viability of the Sea Base for SOF

The ability of MPF(F) platforms to accommodate a JSOTF as an AFSB fits well within the previously modeled MCO and COIN scenarios. Moreover, the MLP, LHA(R), LHD, and T-AKE each offer plenty of open decks and accessibility to interoperate with various special operations surface, air, and undersea platforms. Sea basing SOF offers several tactical and operational advantages in terms of security, surprise, and hosting detailed planning and extensive rehearsals for speed and precision. So the question is not whether MPF(F) ships are effective alternatives to support an AFSB: They are. The questions are, will they be available when a crisis erupts or will combatant commanders have to spend days or weeks negotiating for a base ashore in a nearby country? And if that country agrees to a base, will the JSOTF be neutered by political constraints required to secure an agreement for an FSB ashore?

As acquisition planners use MCO scenarios to determine inventory requirements for MPF(F) platforms, they may want to also consider the expanding role of SOF and irregular warfare. More specifically, considering what are, in our view, faulty basing assumptions underlying USSOCOM's current plans, there may be a reasonable need for further analysis to determine more accurately how to support MCOs and CONOPs effectively. Providing SOF commanders forward staging bases that are readily available will help ensure they can successfully respond to terrorist or other unconventional threats in the future.

---

[9] Robert E. Monroe, "4 ESOS Deployment to Dira Dawa, Ethopia Lessons Learned Report," 4th Expeditionary Special Operations Squadron, USAF, January 26, 2007, not available to the general public.

CHAPTER SIX

Conclusions

Our analysis examined various changes to the planned composition of the MPF(F). Most of the scenarios we examined that vary from the 14-ship program of record MPF(F) squadron involved removing some or all of the air-capable ships—the LHA(R)s and LHD. Additionally, we examined some cases where the a fourth MLP was added, with extra LCACs, in order to examine a situation where only surface connectors were available. Most of this analysis focused on MCO situations, but we did devote some attention to the possibility of the MPF(F) supporting COIN and SOF operations. Finally, our analysis did not examine the ability of the MPF(F) to support joint operations; rather, it concentrated on the ability of a modified MPF(F) to sustain MEBs.

Some key findings of our analysis of various scenarios:

- *Eliminating one LHA(R).* The degradation to logistics throughput resulting from the elimination of one LHA(R) could be offset in all cases by substituting CH-53K helicopters for MV-22 tilt-rotor aircraft; CH-53K helicopters have three times the payload of the MV-22 and, in our scenarios, are just as fast on ingress.[1] The MV-22's higher speed is advantageous in CASEVAC operations, where time is critical and external loads do not limit its speed.

[1] Both helicopters and tilt-rotor aircraft are expected to carry external loads in sustainment operations. The advantage of greater payload weight for internal loads is more than offset by the additional loading time required for external loads. On ingress, aerodynamic constraints imposed by external loads make the MV-22 no faster than the CH-53K. The MV-22 can employ its high speed only on returning from the shore to the sea base.

- *Eliminating both LHA(R)s.* The degradation to logistics throughput resulting from eliminating both LHA(R)s cannot be offset by substituting CH-53K helicopters for MV-22 tilt-rotor aircraft; too few aircraft then remain in the MPF(F) squadron. However, a robust throughput capacity remains for all cases considered using air connectors from the remaining LHD and LCACs from the LHD and the MLPs.
- *Eliminating all large decks.* The elimination of all three large decks (both LHA[R]s and the LHD) in the MPF(F), with sustainment conducted entirely using LCACs from MLPs, leaves a marginal capacity to sustain a single MEB (either in MCO or in COIN operations) with three or four MLPs. However, this option also strips out the MPF(F) squadron's major medical capabilities and forces a reliance on slower aircraft for CASEVAC. Aviation command and control capabilities provided by the LHD would also be lost. Further, the ability of MLPs to work with T-AKEs could be constrained by the relatively small number of helicopters carried for VERTREPs by T-AKEs. Finally, removing the aircraft associated with the large flight decks could impose tactical constraints on the MEB commander. While our analysis showed that, in terms of raw throughput, the LCACs of three or four MLPs could meet the MEB's daily tonnage requirements, we found that significant operational issues could arise without air-capable ships in the force. When sustaining two MEBs in conjunction with an ATF, throughput capacity is marginal without the addition of a fourth MLP. However, the loss of medical and CASEVAC capabilities is less of an issue. The issues of ability of the MLPs to work with T-AKEs and constraints on the MEB commander are also mitigated by the presence of an ATF.
- *LCAC capacity.* The bulk of ship-to-shore throughput capacity for MPF(F) connectors resides with LCACs. The combined total throughput capacity of the 21 LCACs carried by the MPF(F) alone significantly exceeds daily tonnage requirements for the 2015 MEB. Moreover, MCCDC's sustainment plans use the three LCACs carried by the LHD; the use of LCACs cannot be discounted completely. However, as noted below, we uncovered

important issues associated with a heavy reliance upon LCACs for sustainment.

- *Ashore connectors.* Supplies delivered to the shore by LCACs must be moved forward from the beach (or small port) to the USMC or other forces that will consume the supplies. Such movement requires a quantity of trucks and/or aircraft and a reasonably secure area through which they can move. These conditions will, of course, be situationally dependent.
- *T-AKE/MLP interface.* The USMC does not currently envision a direct T-AKE/MLP interface; the offloading of supplies from the T-AKEs is presently limited to aircraft-only sustainment. This concept would shift the burden of lift from LCACs to vertical-lift connectors and so reduce the number of CH-53K and MV-22 sorties available to joint force commanders for purposes other than sustainment.[2] In order to realize the full potential of the MLPs' LCACs, we recommend the Navy and USMC investigate ways that the T-AKEs could interface more closely with the LCACs, either by directly offloading onto the hovercraft themselves or by transloading supplies from T-AKE to MLP, and then into the LCACs.
- *Other LCAC missions.* If the Marine Corps cannot use the full potential of the LCACs, the joint force commander should consider ways to use LCACs for movement ashore and sustainment of other forces. For example, if the MEB does not need, or cannot make use of, the LCACs' throughput potential, the Army could offload personnel, supplies, and equipment onto the MLPs from Army LMSRs for movement ashore via LCAC.
- *Support for COIN.* The MPF(F) sea base, or portions of it, could provide important capabilities to support COIN operations. Although the daily tonnage requirements of a MEB engaged in COIN operations are situationally dependent, they would be lower than the consumption rates envisioned for MCO, especially in terms of ammunition. Therefore, the overall logistics throughput potential of the MPF(F) could easily support a MEB engaged in

---

[2] This topic is explored in some detail in Appendix A.

COIN operations, as well as additional USMC units or elements from the other services. Given the general desire that local forces have a leading role in COIN, the MPF(F) might also be used to support foreign forces. Finally, COIN operations might not require the employment of all the ships of the MPF(F), depending on the size and duration of the mission.

- *Support for SOF.* The MPF(F) could provide a useful base for SOF operations. Even more than in support of COIN operations, support to SOF might require only a portion of the MPF(F). For example, a single MLP or an MLP plus a large flight deck from the MPF(F) might be sufficient to meet the needs of a SOF element, possibly for a protracted period of time.

APPENDIX A

Additional Cases

Two sets of additional cases are explored in this appendix. The first scenarios represent some excursions not included in the main body of this report. The second set of scenarios represents operational issues suggested by MCCDC.

Scenario Excursions

Our treatment of sustaining two MEBs in Chapter Three began with the case in which surface and aerial connectors sustain a near MEB and a further MEB is sustained by air only. As was shown (in Figure 3.9) air-only sustainment of the both MEBs (using the planned aircraft mix) is possible out to a distance of about 70 NM. What, then, if CH-53K helicopters replaced those MV-22 aircraft not reserved for CASEVAC? Figure A.1 shows that with such a replacement, sustainment of the further MEB is possible out to a distance of about 100 NM; it is nearly feasible at distances out to 110 NM. To illustrate, RAND analysts previously investigated the implications of reducing demand for external sustainment, such as that realized by eliminating ground elements' demand for bulk water.[1] Not shown, eliminating the need for bulk water for one of the MEBs would make air-only sustainment possible with CH-53K helicopters replacing those MV-22 aircraft not reserved for CASEVAC. As before, eliminating one LHA(R) from the MPF(F)

[1] Button et al., 2007.

Figure A.1
Air-Only Sustainment of Further MEB Nearly Feasible With CH-53K Helicopters Replacing MV-22 Aircraft

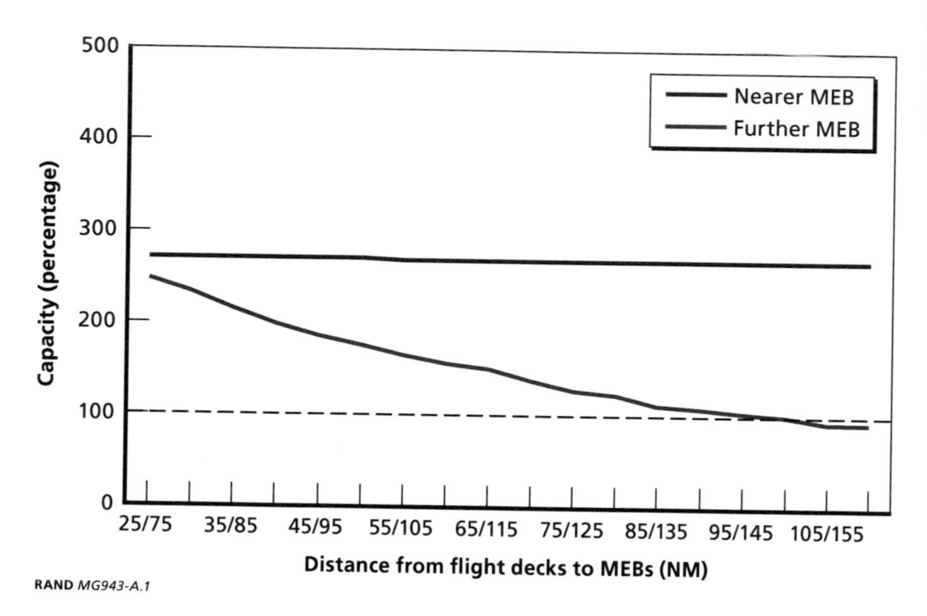

RAND *MG943-A.1*

and replacing MV-22 aircraft with CH-53K helicopters have offsetting effects; the net effect closely resembles Figure 3.9.

In this instance, there is no capability lost with the elimination of an LHA(R) from the MPF(F). There is, however, an opportunity cost in the sense that a capability that could be gained becomes unreachable with the removal of an LHA(R) from the MPF(F).

Separate Operational Concepts

MCCDC personnel suggested the possibility that in some cases, perhaps 30 percent of sorties by MPF(F) aerial connectors would be for purposes other than logistic support. Model results indicate (not surprisingly) that devoting aircraft sorties for purposes other than logistic support somewhat reduces support capacity. Sustainment from 110 NM, previously marginal, is not feasible with 30 percent of (CH-53K and

MV-22) aerial connector sorties devoted to other purposes; as shown in Figure A.2, the feasible limit to sustainment is now about 90 NM.[2]

The baseline case with 70 percent of MPF(F) aerial connectors is then a more meaningful test. Results for this case and corresponding cases with one or both LHA(R)s removed are shown in Figure A.3.

MCCDC personnel also suggested the possibility that LCACs could not participate in moving dry stores and ammunition from sea base T-AKEs ashore. Rather than using VERTREP or CONREP to move dry stores and ammunition to the MLP for loading onto LCACs, they would only be transported ashore by CH-53K and MV-22 aircraft.

Figure A.2
Single MEB Supported by All or 70 Percent of MPF(F) Aerial Connectors Plus LHD LCACs

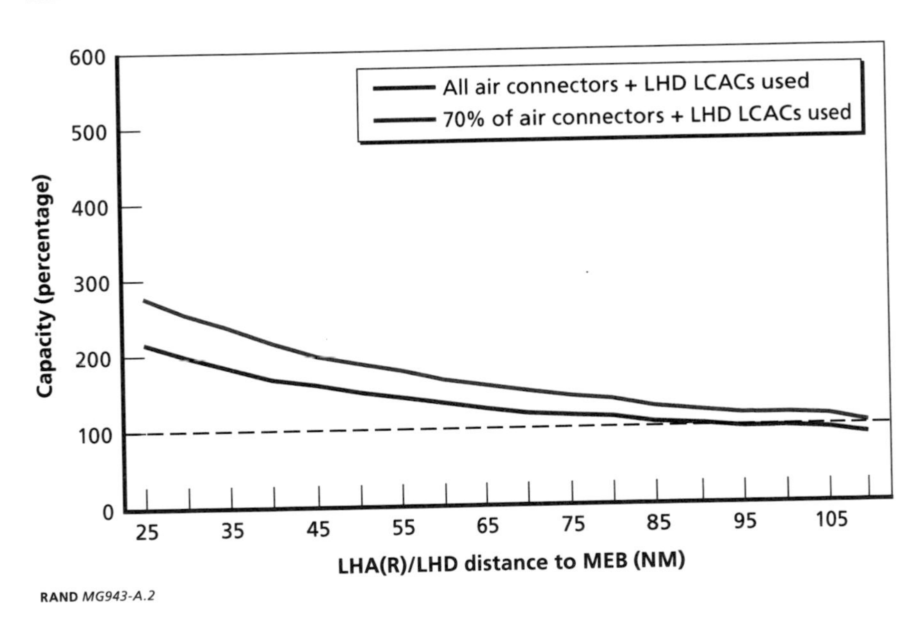

RAND *MG943-A.2*

2 The results shown in Figure A.2 were derived by removing 30 percent of all CH-53K and MV-22 logistics aircraft from the MPF(F) (leaving alone aircraft reserved for CASEVAC). With reduced competition for deck spots for launching and recovering aircraft, this had the effect of increasing individual aircraft efficiency so that aircraft throughput was reduced by 25 percent. Our results are therefore somewhat optimistic.

Figure A.3
Single MEB Supported by All or 70 Percent of MPF(F) Aerial Connectors

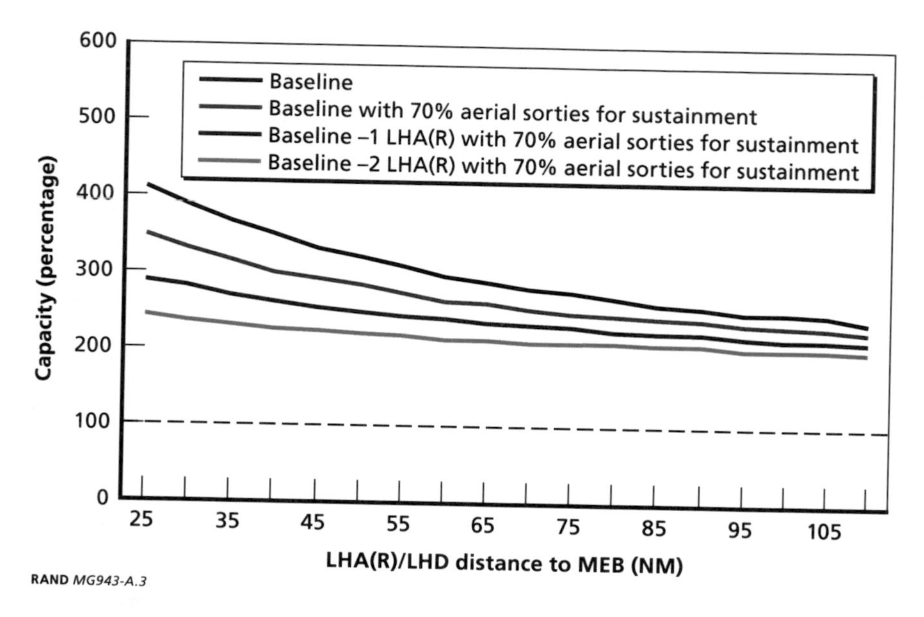

RAND *MG943-A.3*

This situation might reflect conditions when movement by LCAC is not acceptable.

This issue was addressed by adding the simulation option of not allowing LCACs to carry ammunition from T-AKEs.[3] We found that throughput capacity in support for a single MEB when LCACs cannot be used to transport ammunition is nearly identical to throughput capacity in the baseline case (see Figure A.4). With connectors not as well matched to payloads, there is a slight (several percent) degradation in throughput capacity. Similar patterns (not shown) emerge when one or both LHA(R)s are removed; we did not see the need to evaluate throughput using CH-53K helicopters in place of MV-22 aircraft. The restriction that only aerial connectors can transport dry stores and

---

[3] Ammunition is the main component of the dry stores and ammunition taken from T-AKEs. The percentage of dry stores moved from LMSRs could not be ascertained, so in this scenario limitations were placed solely on ammunition movement. Ammunition movement from the LHD was not restricted. In our simulations, about three LCAC sorties per day from the LHD carried ammunition.

Figure A.4
Single MEB With Ammunition Movement by Air Connectors Only

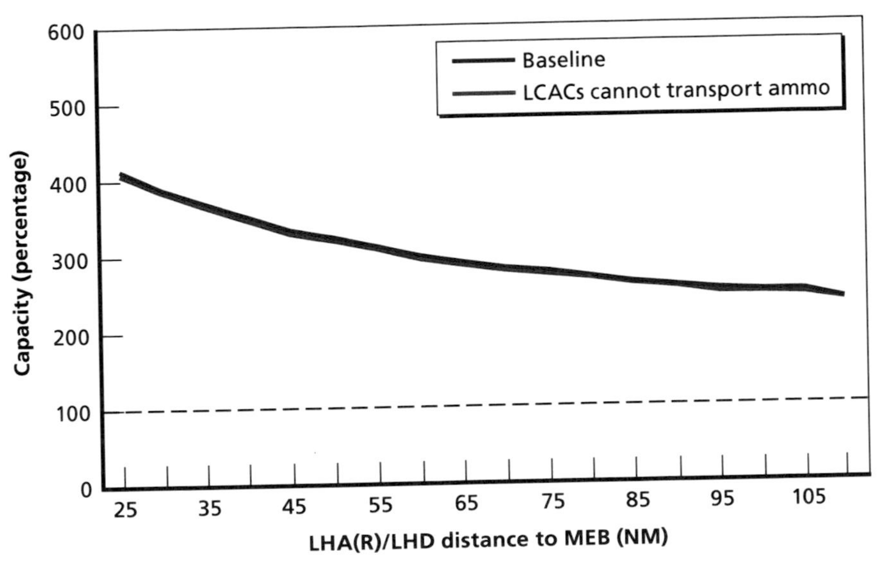

RAND *MG943-A.4*

ammunition therefore has a negligible effect on throughput capacity. Of greater significance, this restriction shifts the lift burden from LCACs to vertical lift connectors.[4] The net effects then are to (1) decrease the number of aircraft sorties available to the joint force commander for purposes other than sustainment and (2) increase LCAC sorties similarly available. As MV-22 sorties may be more valuable than LCAC sorties to joint force commanders, this suggests value in improving VERTREP or CONREP capabilities between T-AKEs and MLPs.

---

4 Specifically, the simulation increased the number of MV-22 sorties carrying ammunition by an average of 35 sorties per day.

APPENDIX B

Maritime Prepositioning Force (Future) Description

At the time of this analysis, the planned 14-ship MPF(F) squadron will consist of

- two LHA(R) large-deck amphibious assault ships
- one modified LHD large-deck amphibious ship
- three *Lewis and Clark* (T-AKE) cargo ships
- three modified LMSR sealift ships
- three mobile landing platform (MLP) ships each capable of operating six landing craft, air cushioned (LCAC) surface connectors
- two legacy dense-pack MPF ships taken from existing squadrons.

LHA(R)s and LHDs have large flight decks and hangar decks for embarking and operating helicopters and MV-22 tilt-rotor aircraft. This appendix describes MPF(F) ships other than the existing MPF ships, which are not relevant to this study.

LHA(R) and LHD

The notional LHA(R) Flight 0 large-deck amphibious ship will be a modified version of the LHD-8 amphibious assault ship. Designated LHA-6, it is notable for its lack of a well deck—meaning that it cannot operate LCACs or landing craft utility (LCU) boats.[1] It will have nine

---

[1] The LCU is a surface connector used to transport personnel and equipment to the shore. LCUs can transport tracked or wheeled vehicles and troops from amphibious assault ships to

aviation landing spots, six on the port side. The MPF(F) LHA(R) is distinguished from an ESG LHA(R) in its simplified command and control system and lack of active defense systems. Future LHA(R)s will also be developed in ESG and MPF(F) versions; relative to the ESG LHA(R), the MPF(F) LHA(R) will have a simplified command and control system and lack active defense systems. LHA(R)s and LHDs also provide medical capabilities: With six operating rooms, 17 intensive care unit beds, and 60 overflow beds, LHDs have the greatest medical capability of any amphibious platform in operation. LHDs can operate three LCAC-equivalent connectors and have nine aviation landing spots, seven on the port side. The MPF(F) LHD will be a decommissioned LHD from the fleet modified for MPF(F).

MPF(F) LHA(R) and LHD vessels are to collectively carry a 2015 MEB air combat element, which include 48 MV-22, 20 CH-53K, and 18 AH-1 helicopters. Each aviation ship is to carry two SH-60 helicopters.[2] Both the LHA(R) and the LHD will store 400,000 gallons of water and produce 200,000 gallons of water per day.

A current LHD, the USS *Bataan* (LHD-5), is shown with MV-22 aircraft spotted in Figure B.1.

T-AKE Cargo Ships

The T-AKE is an underway replenishment naval vessel of the Combat Logistics Force. It has two multi-purpose cargo holds, capable of selective offload, for dry stores and/or ammunition.[3] It has additional holds for freeze, chill, and/or dry stores, and three specialty and spare parts cargo holds. Its capacity for dry cargo is up to 886,963 lbs. It can carry up to 24,959 barrels of cargo fuel.[4]

The T-AKE has a single day/night capable vertical replenishment (VERTREP) station. It also has three dry cargo and one liquid cargo

---

beachheads or piers.

[2] The Secretary of the Navy approved this squadron in May 2005.

[3] The bulk of Joint Strike Fighter munitions will be carried on T-AKEs.

[4] U.S. Navy, "U.S. Navy Fact File: Dry Cargo/Ammunition Ships—T-AKE," web page.

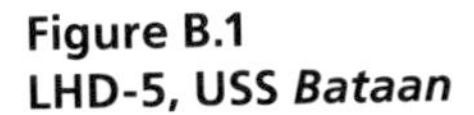

Figure B.1
LHD-5, USS *Bataan*

RAND *MG943-B.1*

SOURCE: U.S. Navy, V-22 program website.

connected replenishment (CONREP) stations on each side. It can simultaneously operate five CONREP stations or three CONREP stations while conducting VERTREP operations.

The design speed of the T-AKE is 20 knots. The lead ship of the class, the USNS *Lewis and Clark*, is shown in Figure B.2 below.

LMSR Cargo Ships

The MPF(F) LMSR will have about 202,000 square feet of cargo space, will have two or four aircraft operating spots, and will berth about 850 personnel. Its design speed is 20 knots. It will store 33,500 gallons of water and will have the capacity to produce 24,000 gallons of water per day. The Military Sealift Command's newest LMSR, the USNS *Soderman*, is shown in Figure B.3.

Figure B.2
T-AKE-1, USNS *Lewis and Clark*

RAND *MG943-B.2*

SOURCE: U.S. Navy, Military Sealift Command Ship Inventory.

Figure B.4 illustrates an MPF(F) LMSR (on the right) transferring a vehicle to a notional MLP.

Mobile Landing Platform

The Mobile Landing Platform (MLP) (shown conceptually in Figure B.5) will be a clean sheet design leveraging existing float-on/float-off technology. It will carry six LCAC-equivalent connectors, with one aircraft vertical replenishment spot capable of operating a heavy lift helicopter for VERTREPs. It will also have five UNREP sta-

Figure B.3
T-AKR-317, USNS *Soderman*

RAND *MG943-B.3*

SOURCE: U.S. Navy, Military Sealift Command Ship Inventory.

tions.[5] It will have accommodations for 922 personnel, including crew. Its design speed will be 20 knots.

---

5 The MLP will have one replenishment-at-sea station (receive only), two diesel fuel (marine) stations (receive only), and two JP-5 stations (one send and one receive). Source: Program Executive Office (PEO) Ships.

Figure B.4
Vehicle Transfer Between MPF(F) LMSR and Conceptual Mobile Landing Platform

RAND *MG943-B.4*

SOURCE: PEO Ships.

Figure B.5
Conceptual MLP

RAND *MG943-B.5*

SOURCE: PEO Ships.

APPENDIX C

MPF(F) MEB Sustainment Requirements

The MPF(F) MEB, or 2015 MEB, was developed by the Marine Corps for operation from MPF(F) ships. The 2015 MEB has 14,484 personnel, organized into a Sea Base Echelon (with 8,397 personnel), a Forward Base Echelon (with 1,907 personnel), and a Sustained Operations Ashore Echelon (with 4,180 personnel). The Sea Base Echelon (SBE) has a Sea Base Maneuver Element (SBME) with 4,989 personnel and a Sea Base Support Element (SBSE) with 3,408 personnel. We evaluate the SBME here. The SBSE initially supports the SBME from the sea base; later much of it will move ashore to better support the SBME. The Sustained Operations Ashore Echelon normally operates from the continental United States.

Marine Corps sustainment is grouped, as in the CDD analysis, as ammunition, dry stores, and bulk water and petroleum, oil and lubricants (POL) per day. Ammunition and dry stores are measured in short tons per day. Bulk water and POL are measured in gallons per day.

MEB consumption data for the cases examined in this study are shown below. Number of personnel drives the consumption of water and dry stores; this explains the identical entries seen in Table C.1.

Table C.1
SBME Daily Sustainment Requirements

	Water (ST)	POL (ST)	Ammunition (ST)	Dry Stores (ST)
MCO	170	444	302	51
COIN-Low	170	222	30	51
COIN-High	170	399	151	51

APPENDIX D

Model Description

Overview

RAND analysts originally developed the JSLM in 2006 and 2007 by reverse engineering analytic tools used by MCCDC for the MPF(F) Initial Capabilities Document (ICD) analysis. JSLM was developed as a tool to ascertain the feasibility of simultaneously sustaining USMC and Army elements ashore from a sea base, or the feasibility of moving an Army element in a reasonable period while sustaining a Marine Corps ground element already ashore.

JSLM is an object-oriented simulation model. To illustrate the object-oriented concept, amphibious assault ships (LHA[R]s and LHDs) and rotary-wing aircraft (CH-53s and MV-22s) are objects in the model acting on each other in well-defined ways. They are capable of sending and receiving messages and processing data; they can be viewed as independent actors with defined roles and responsibilities. Similarly, MLPs and LCAC landing craft are objects that interact with each other within the model.[1] Operational friction stems from these interactions. For example, an aircraft returning to the sea base can be advised that there is no landing spot available for it and that it must loiter until one becomes available. Or a fueled and loaded aircraft must wait until another has cleared the flight deck before it can take off. Aircraft must conclude operations before the flight window of an

[1] The object-oriented paradigm is supported using the high-level programming language C++, which has features facilitating object-oriented programming. JSLM has about 3,000 lines of source code.

amphibious ship ends. A limited number of LCACs can conduct cyclic operations simultaneously on an MLP; others may have to wait their turn. The duration of a single LCAC, CH-53, or MV-22 sortie is easy to calculate; it is the friction in the system that drives the problem.

Operationally, JSLM can be described as a time-stepped deterministic simulation. The time step is one minute. Such a small time step is needed to capture such brief events as the delay experienced by one aircraft ready for take off as another aircraft takes off. Because it is deterministic, the model cannot directly treat such issues as equipment failure or aircraft losses to antiaircraft systems. The model can be used to treat them indirectly through such techniques as reducing the number of operational LCACs or aircraft at the outset.

The original version of JSLM was minimally modified for this study with changes limited to such areas as making the number of MLPs a variable. Adding the capability for the MPF(F) LHD (as well as MLPs) to operate LCACs was the single biggest model change for this analysis.

We modeled logistic sustainment for an Army brigade and a MEB using this modified version of JSLM. In order to treat sustainment for two MEBs, Army brigade combat team (BCT) sustainment requirements were replaced with those of the MEB. The model "thought" it was sustaining a BCT and a MEB. In order to treat sustainment for just one MEB, Army BCT sustainment requirements were zeroed—the MEB provided the only sustainment demand signal. Further, the BCT and the MEB were collocated so that there was no payload penalty for sustainment efforts to the phantom BCT.

Model Data

This study capitalized upon recent sea basing studies conducted by MCCDC and the Center for Naval Analyses (CNA). The MCCDC and CNA analyses examined sustainment using only aircraft. CH-53K and MV-22 parameters used for this analysis were provided by MCCDC. MLP characteristics are based on a PEO Ships MLP notional design and the MV *American Cormorant*. LCAC data are from the *MAGTF Planner's Reference Manual* as updated for service life extension.[2]

MEB sustainment requirements were provided for this study by MCCDC.

Testing

Two opportunities to verify JSLM against MCCDC modeling results have presented themselves. In the 2007 study for which JSLM was initially developed, MCCDC personnel stated that air-only sustainment for the 2006 version of the MPF(F) MEB was barely possible with at a distance of 110 NM from the Sea Base to the sustained MEB. Our original study confirmed that finding. For this study, MMCDC personnel found that for the current version of the MPF(F) MEB, air-only sustainment at a distance of 110 NM was not possible but was feasible with three LCACs from the MPF(F) LHD. As was shown in Figure 3.2 (repeated below as Figure D.1) JSLM results are in close agreement with the MCCDC finding.

---

2 Marine Corps Combat Development Command MSTP Center, *MAGTF Planner's Reference Manual*, MSTP Pamphlet 5-0.3, Quantico, Va., August 2006.

Figure D.1
Lift Capacity Using LHA(R)/LHD Assets in MEB Sustainment

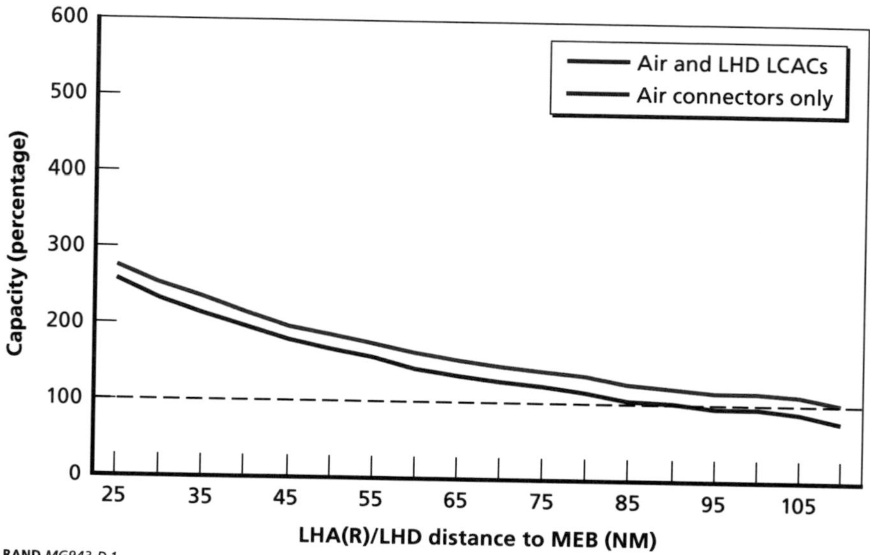

RAND *MG943-D.1*

Bibliography

Button, Robert W., John Gordon IV, Jessie Riposo, Irv Blickstein, and Peter A. Wilson, *Warfighting and Logistic Support of Joint Forces from the Joint Sea Base*, Santa Monica, Calif.: RAND Corporation, MG-649-NAVY, 2007. As of March 11, 2010:
http://www.rand.org/pubs/monographs/MG649/

"Capability Development Document of the Maritime Prepositioning Force (Future) [MPF(F)] Squadron: Increment One Mobile Landing Platforms (MLP) and Auxiliary Dry Cargo/Ammunition Ships (T-AKE)," November 16, 2007.

Clark, ADM Vern, CNO, and Gen. Michael Hagee, CMC, "Naval Operating Concept for Joint Operations," Department of the Navy, September 2003.

Congressional Budget Office, "Sea Basing and Alternatives for Deploying and Sustaining Ground Combat Forces," 2007.

Defense Science Board, *Enabling Sea Basing Capabilities*, August 2003.

Defense Science Board Task Force on Mobility, *Final Report*, Washington, D.C., Office of the Under Secretary of Defense for Acquisition, Technology, and Logistics, September 2005.

Department of Defense, "Seabasing Joint Integrating Concept Version 1.0," Washington, D.C., August 2005.

Futcher, LCDR Frank, "Seabasing Logistics Concept of Operations, OPNAV N42," briefing, May 2005.

General Dynamics, "Lewis and Clark (T-AKE 1) Class Dry Cargo/Ammunition Fact Sheet," January 2007.

Hagee, Gen. Michael, CMC, "Concepts and Programs," Department of the Navy, 2006.

Headquarters United States Marine Corps, "MCRP 3-31.1A: Employment of Landing Craft Air Cushion (LCAC)," Washington, D.C., 1997.

———, "The STOM Concept of Operations (STOM CONOPS)," Washington, D.C., draft as of April 2003.

———, *Prepositioning Programs Handbook*, PCN 50100234000, Washington, D.C., March 2005.

———, "Amphibious Requirements: USN and USMC Warfighter Talks," briefing, February 2, 2007.

Henning, CDR Mark, USN, "U.S. Navy Transformation: Sea Basing as Sea Power 21 Prototype," USAWC Strategy Research Project, U.S. Army War College, March 2005.

Joint Chiefs of Staff, *Joint Tactics, Techniques and Procedures for Landing Force Operations*, Joint Publication 3-02.1, May 11, 2004.

———, *Joint Special Operations*, Joint Publication 3-05, 2006.

———, *Amphibious Operations*, Joint Publication 3-02, August 10, 2009.

Kaskin, Jonathan, "The Challenge of Seabasing Logistics," OPNAV N42, briefing, February 17, 2005.

Kaskin, Jonathan, "Seabasing Logistics CONOPs," OPNAV N42, briefing to NDIA 10th Annual Expeditionary Warfare Conference, October 2005.

Kohut, Andrew, "America's Image in the World: Findings from the Pew Global Attitudes Project," testimony before the Committee on Foreign Affairs, U.S. House of Representatives, March 14, 2007.

Kurinovich, Mary Ann, and Michael W. Smith, "Sustainment of MPF(F) Squadron by T-AKEs," CAB D0014916.A2/Final, Center for Naval Analyses, October 2006.

Marine Corps Combat Development Command, *Experimental Marine Expedition Brigade Planner's Reference Guide*, Quantico, Va., 2002.

Marine Corps Combat Development Command, Mission Area Analysis Branch, "MPF(F) CDD Analysis: Results for Seabasing Capabilities," briefing, March 23, 2006.

———, "Surface Assault Connector Requirements Analysis Update: Overview to Inform Seabasing Capabilities Study," briefing, April 13, 2006.

Marine Corps Combat Development Command, MSTP Center, *MAGTF Planner's Reference Manual*, MSTP Pamphlet 5-0.3, Quantico, Va., August 2006.

McCarthy, VADM Justin, Director, OPNAV N4, "Seabasing Logistics," presentation to the NDIA 10th Annual Expeditionary Warfare Conference, briefing, October 2005.

McRaven, William H., *Spec Ops: Case Studies in Special Operations Warfare Theory and Practice*, Novato, Calif.: Presidio Press, 1995.

Monroe, Robert E., "4 ESOS Deployment to Dira Dawa, Ethopia Lessons Learned Report," 4th Expeditionary Special Operations Squadron, USAF, January 26, 2007, not available to the general public.

Naval Research Advisory Committee, *Sea Basing*, NRAC 05-2, March 2005.

Naval Studies Board, National Research Council, *Sea Basing: Ensuring Joint Force Access from the Sea*, Washington, D.C.: The National Academies Press, 2005.

O'Rourke, Ronald, *Navy-Marine Corps Amphibious and Maritime Prepositioning Ship Programs: Background and Oversight Issues for Congress,* Congressional Research Service RL32513, updated July 10, 2007.

Perry, Walter L., and John Gordon IV, *Analytic Support to Intelligence in Counterinsurgencies*, Santa Monica, Calif.: RAND Corporation, MG-682-OSD, 2008. As of January 12, 2010:
http://www.rand.org/pubs/monographs/MG682/

Robbins, Darron L., and Michael W. Smith, "Resupplying Forces Ashore Using Sea-Based Aircraft," CAB D0014746.A2/Final, Center for Naval Analyses, September 2006.

U.S. Joint Chiefs of Staff, *Joint Special Operations*, Joint Publication 3-05, 2006.

U.S. Navy, "U.S. Navy Fact File: Dry Cargo/Ammunition Ships—T-AKE," web page. As of October 5, 2009:
http://www.navy.mil/navydata/fact_display.asp?cid=4400&tid=500&ct=4

U.S. Special Operations Command, *Capstone Concept for Special Operations*, MacDill Air Force Base, Fla., 2006.

———, *United States Special Operations Command History, 1987–2007*, 2007a.

———, *USSOCOM Posture Statement 2007*, MacDill Air Force Base, Fla., 2007b.

Work, Robert, *Thinking About Seabasing: All Ahead, Slow*, Washington, D.C.: Center for Strategic and Budgetary Assessments, 2006.

Yopp, J. J., "Future Seabasing Technology Analysis: Logistics Systems," CRM D0014262.A2/Final, Center for Naval Analyses, August 2006.